Advances in
THE STUDY OF BEHAVIOR
VOLUME 3

Contributors to This Volume

C. G. BEER

LAWRENCE V. HARPER

D. J. McFARLAND

Y. SPENCER-BOOTH

JANE van LAWICK-GOODALL

Advances in
THE STUDY OF
BEHAVIOR

Edited by

Daniel S. Lehrman
Institute of Animal Behavior
Rutgers University
Newark, New Jersey

Robert A. Hinde
Sub-Department of Animal Behaviour
Cambridge University
Cambridge, England

Evelyn Shaw
American Museum of Natural History
New York, New York

——————————— VOLUME 3 ———————————

ACADEMIC PRESS New York and London, 1970

591.51
A224
v.3

ACADEMIC PRESS, INC.
111 Fifth Avenue, New York, New York 10003

United Kingdom Edition published by
ACADEMIC PRESS, INC. (LONDON) LTD.
Berkeley Square House, London W1X 6BA

LIBRARY OF CONGRESS CATALOG CARD NUMBER: 64-8031

PRINTED IN THE UNITED STATES OF AMERICA

Contents

Behavioral Aspects of Homeostasis
D. J. McFARLAND

Individual Recognition of Voice in the Social Behavior of Birds
C. G. BEER

Ontogenetic and Phylogenetic Functions of the Parent-Offspring Relationship in Mammals
LAWRENCE V. HARPER

The Relationships between Mammalian Young and Conspecifics
Other Than Mothers and Peers: A Review

Y. SPENCER-BOOTH

Tool-Using in Primates and Other Vertebrates

JANE van LAWICK-GOODALL

List of Contributors

C. G. BEER, *Rutgers University, Newark, New Jersey*

LAWRENCE V. HARPER,* *Child Research Branch, National Institute of Mental Health, Bethesda, Maryland*

D. J. McFARLAND, *Department of Experimental Psychology, Oxford University, Oxford, England*

Y. SPENCER-BOOTH, *Sub-Department of Animal Behaviour, Madingley, Cambridge, England*

JANE van LAWICK-GOODALL, *Gombe Stream Research Centre, Kigoma, Tanzania*

*Present address: Department of Applied Behavioral Sciences, University of California, Davis, California

Preface

The study of animal behavior is attracting the attention of ever-increasing numbers of zoologists and comparative psychologists in all parts of the world, and is becoming increasingly important to students of human behavior in the psychiatric, psychological, and allied professions. Widening circles of workers, from a variety of backgrounds, carry out descriptive and experimental studies of behavior under natural conditions, laboratory studies of the organization of behavior, analyses of neural and hormonal mechanisms of behavior, and studies of the development, genetics, and evolution of behavior, using both animal and human subjects. The aim of *Advances in the Study of Behavior* is to provide for workers on all aspects of behavior an opportunity to present an account of recent progress in their particular fields for the benefit of other students of behavior. It is our intention to encourage a variety of critical reviews, including intensive factual reviews of recent work, reformulations of persistent problems, and historical and theoretical essays, all oriented toward the facilitation of current and future progress. *Advances in the Study of Behavior* is offered as a contribution to the development of cooperation and communication among scientists in our field.

Contents of Previous Volumes

Advances in
THE STUDY OF BEHAVIOR
VOLUME 3

Behavioral Aspects of Homeostasis[1]

D. J. McFarland

DEPARTMENT OF EXPERIMENTAL PSYCHOLOGY
OXFORD UNIVERSITY
OXFORD, ENGLAND

"La fixité du milieu intérieur est la condition de la vie libre." Claude Bernard's celebrated maxim provides a keystone for the study of biological regulating mechanisms. The idea of regulation was implicit in Claude Bernard's concept of the constancy of the internal environment, though the explicit postulate of the role of nervous and chemical feedback mechanisms in the maintenance of physiological steady state was formulated later, and embodied in Walter Cannon's concept of homeostasis. It is now well recognized that there are important behavioral links in homeostatic mechanisms, and that much behavior is ultimately concerned with the maintenance of an optimal milieu intérieur. The recent application of control theory to the analysis of homeostatic mechanisms suggests that behavior can also be analyzed in these terms.

I. CONTROL THEORY AND BEHAVIOR

Although it is a truism to say that behavior is a function of time, this aspect of behavior is by no means the only, or even the most common, focal point for behavior study. In studying genetic or evolutionary as-

[1]This paper was written as part of a project on behavioral systems analysis financed by the Science Research Council. The author is grateful to Dr. James Dewson III and to Professor L. W. Weiskrantz for their helpful criticism of the manuscript.

1

pects of behavior, for instance, the way in which behavior changes from one generation to the next is of prime importance. Similarly, in learning studies, behavior is generally considered as a function of trials, or some other measure of experience. In studying behavior as a function of time, one is primarily concerned with the performance of particular behavior patterns, their orientation, goal directiveness, and motivation. One is concerned with the manner in which such behavior is controlled by internal and environmental stimuli. Control theory is particularly appropriate to this type of study, and although it could in principle be applied to other aspects of behavior, its present application is restricted to the study of motivation.

There are two basic approaches to the description of a system in terms of control theory: the synthetic approach, and the analytic approach. The synthetic approach consists in putting together elements or subsystems of known input-output characteristics, as described in Fig. 2. The functional relationships between the elements prescribe the behavior of a hypothetical or abstract system, the validity of which is tested experimentally. This approach is of great value in integrating the interactions between components of large and complex systems, where the components have previously been studied in isolation. Recent applications of such methods to systems controlling body temperature (Crosbie et al., 1961), respiration (Milhorn, 1966), and cardiovascular mechanisms (Grodins, 1959) herald a new integrative approach to the physiology of homeostatic mechanisms.

The analytic approach consists in the investigation of the input-output relationships of the system. Inputs of known waveform are applied and correlated with the measured outputs. For simple systems transient analysis is adequate, and an example of this will be described later. For higher-order systems frequency analysis is appropriate. This method has been used extensively in the transfer function determination of sensory receptors. A harmonic stimulus is applied to the receptor, and its response is measured, either behaviorally (e.g., Thorson, 1966), or neurophysiologically (e.g., Pringle and Wilson, 1952; Chapman and Smith, 1963). The input-output amplitude-ratio and phase-shift can be determined as a function of frequency, and the transfer function of the system calculated on this basis. Stark and Young (1964) have pointed out some of the dangers of applying such methods indiscriminately to biological systems. In particular, slow-adaptation (Stark, 1959), pre-cognitive input prediction (Stark et al., 1962), and task adaptation (Young et al., 1963) have been demonstrated in quasi-behavioral situations. Although the presence of such phenomena can sometimes invalidate certain methods of investigation, a large battery of systems analysis techniques is now available, and an alternative method of analysis can usually be found. Of particular interest to biologists are the recently developed techniques of random-data analysis (e.g., Bendat and Piersol, 1966; Korn, 1966), which are especially suited to the analysis of systems subject to random variations in the values of important variables. Such variations are common in behavioral measurements, but this does not mean that deterministic methods of investigation are not possible. Indeed, even at an elementary level, it is sometimes possible to use a noisy input as a tool in systems analysis. McFarland (1967) used variations in ambient temperature to demonstrate the phase relationships between feeding and drinking in pigeons.

To describe a motivational system in terms of control theory, it is necessary to define or isolate a system to be studied. A system may be defined as *an ordered arrangement of physical or abstract entities;* and a motivational system can be envisaged as a network of functionally related variables, relevant to the behavior in question. The decision as to what variables are relevant may have to be an arbitrary one, based on the theoretical attitude of the investigator. For example, the description of a motivational system looked at from a "general drive" viewpoint will be different from that constructed on a "specific drive" hypothesis (Hinde, 1966; McFarland, 1966b). This type of distinction is largely an empirical matter, however, and there is no difficulty of principle here.

In the case of homeostatic behavior, the problem is somewhat simplified as behavioral changes can always be measured in terms of their ultimate effect on the animal's internal environment. In other words, such behavior can be identified as belonging to a clearly defined feedback loop. In a homeostatic system the behavioral loop will generally be in parallel with a number of purely physiological loops. For example, the loop subserving water ingestion is essentially parallel to the pituitary-antidiuretic loop; both acting to maintain water balance. For this reason it is essential to study the behavioral element of a homeostatic system in conjunction with the physiological elements; at least during the early stages of the analysis during which the general nature of the system is outlined in terms of control theory. The weighting given to each of the parallel components will vary from one homeostatic system to the next. Thus the behavioral element carries considerable weight in the system controlling water balance, but relatively little in that regulating body temperature. However, before going into details of homeostatic systems, it will be useful to outline the nature of control theory in general.

II. An Outline of Control Theory

Control theory is a discipline that cuts across traditional boundaries

between fields of scientific study, and uses the same language for systems of different hardware, whether physical or biological. Control
theory can also act as a vehicle for analogy between physical and behavioral systems. For example, in physical systems, *energy is defined as
the capacity for doing work*, and *work* as the *quantitative measure of the act of
changing the state of a system*. Thus the concept of energy follows from the
concept of state, which can be defined mathematically in a physical
system, and, by application of control theory, in a behavioral system.
Now if the concept of motivational state can be related to physical state
by mathematical analogy, then the concept of energy should also apply
to motivational systems; and such an analogy might have important
philosophical implications.

A. Dynamic Analogies

Analogies between various laws of nature have proved important in
the development of the physical sciences, particularly in the application
of familiar concepts to fresh fields of study, as instanced by the work of
Georg Simon Ohm (1787–1854), who looked upon the flow of electric
current in a wire as analogous to the flow of heat along a conductor. In
this section the possibility of extending physical analogies to the behavioral sciences is explored.

Two systems can be said to be analogous when their behavior, defined
by an equation, is identical. As an example, consider the mechanical
system portrayed in Fig. 1a. A mass is suspended in a bucket of water,
from a spring attached to a fixed beam. Applying Newton's law — force
equals mass times acceleration — we can write

$$f_m = M \frac{du}{dt} \tag{1}$$

The force needed to push the mass through the water is proportional
to the velocity of the mass. Thus our second equation is

$$f_b = Bu \tag{2}$$

where B is the constant of proportionality, called the friction. The third
equation relates the rate of change of displacement of the spring (i.e.,
the spring velocity) to the applied force

$$f_k = \frac{1}{K} \int u \, dt \tag{3}$$

where K is a constant determining the compliance of the spring. Suppose that no external force is applied after the mass is initially set in

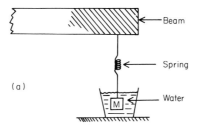

(a)

(b)

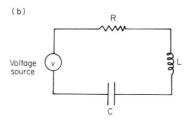

FIG. 1. Analogous mechanical (a) and electrical (b) systems.

motion: what will be the behavior of the system? According to d'Alembert's principle the algebraic sum of the forces applied to a body is zero. In the present case

$$f_m + f_b + f_k = 0 \qquad (4)$$

Substituting Eqs. (1), (2), and (3), the behavior of the system can be described in a single equation,

$$M\frac{du}{dt} + Bu + \frac{1}{K}\int u\,dt = 0 \qquad (5)$$

Now, consider the electrical circuit portrayed in Fig. 1b. A resistance, inductance, and a capacitance are wired in series with a voltage source. The voltage across the inductance is given by the equation

$$v_1 = L\frac{di}{dt} \qquad (6)$$

the voltage across the resistance is given by Ohm's law

$$v_r = Ri \qquad (7)$$

and the voltage across the capacitance is given by the equation

$$v_c = \frac{1}{C} \int i \, dt \tag{8}$$

According to Kirchoff's voltage law, the algebraic sum of all the voltages around a closed loop is zero. Thus if the source is suddenly dropped to zero, the behavior of the system can be described by the equation,

$$v_1 + v_r + v_c = 0 \tag{9}$$

Substitution of Eqs. (6), (7), and (8) yields the following equation:

$$L\frac{di}{dt} + Ri + \frac{1}{C} \int i \, dt = 0 \tag{10}$$

By inspection, Eqs. (5) and (10) can be seen to be similar. In fact, if we substitute u for i, M for L, B for R, and K for C, they become identical. We can justifiably say that the two systems portrayed in Fig. 1 are analogous, and we can draw up a table of analogous variables and parameters, such as Table I. Analogies such as this are well known in the phys-

TABLE I
ANALOGOUS VARIABLES AND PARAMETERS

	Mechanical	Electrical	General
Rate variables	Force	Voltage	Effort
	f	v	e
	Velocity	Current	Flow
	u	i	f
State variables	Momentum	Flux-linkage	Momentum
	p	λ	p
	Displacement	Charge	Displacement
	x	q	h
Parameters	Mass	Inductance	Inductance
	M	L	L
	Compliance	Capacitance	Capacitance
	K	C	C
	Friction	Resistance	Resistance
	B	R	R
Power	$P = fu$	$P = vi$	$P = ef$
Potential energy	$E_p = \frac{1}{2}Kf^2$	$E_p = \frac{1}{2}Cv^2$	$E_p = \frac{1}{2}Ce^2$
Kinetic energy	$E_k = \frac{1}{2}Mu^2$	$E_k = \frac{1}{2}Li^2$	$E_k = \frac{1}{2}Lf^2$

ical sciences and have been extended to acoustic, hydraulic, and thermal systems. Before considering how such analogies might be applied to the behavioral sciences, we must develop our description of physical systems a little further.

The existance of analogies enables us to define *generalized* system variables and parameters. Thus, as illustrated in Table I, the analogy between force and voltage leads to the concept of a generalized *effort* variable e. Similarly, the generalized *flow* variable f comes from the velocity-current analogy. In addition to these two *rate* variables, we can define two generalized *state* variables, where d/dt (state variable) = rate variable. Thus generalized *displacement* h is the integral of the flow variable, and is analogous to mechanical displacement and to electrical charge. Similarly, the generalized *momentum* variable p, the integral of the effort variable, is analogous to mechanical momentum and to electrical flux-linkage.

The *state* of a system can be defined by those variables which would have measureable values if time were "frozen." Clearly, rates can only be measured by reference to the passage of time, so that true rate variables would have zero value if time were frozen. The values of state variables, on the other hand, can be measured at any instant of time. Rate variables are involved in the process of changing the state of a system, the measure of which is defined as *work*. *Power* is defined as the rate of doing work, and is the product of the two rate variables. Thus in terms of the generalized rate variables, power $P = ef$. *Energy*, defined as the capacity for doing work, is the time integral of power, viz. $P = dE/dt$. In practice, flow is always opposed by a resistance, which is overcome by applying a suitable effort. In this process energy is dissipated, or lost from the system to the environment. Thus a generalized *resistance* R can be defined in terms of the generalized rate variables: $e = Rf$. The generalized energy storage elements are the generalized *capacitance* C, measured in terms of displacement per unit effort (thus $h = Ce$ and $e = 1/C \int f dt$); and the generalized *inductance* L with units of momentum per unit flow (i.e., $p = Lf$ and $f = 1/L \int e dt$). The generalized capacitance stores *potential* energy when its effort level is raised by influx of a flow variable. Thus we can write displacement $h = \int f dt$, potential energy

$$E_P = \int_0^h e \, dh = \frac{h^2}{2C} = \frac{Ce^2}{2} \qquad (11)$$

Similarly, the generalized inductance stores *kinetic* energy in accordance with the following equation:

$$\text{momentum } p = \int e \, dt, \quad \text{kinetic energy } E_k = \int_0^p f \, dp = \frac{p^2}{2L} = \frac{Lf^2}{2} \quad (12)$$

In terms of these generalized variables and parameters, which are summarized in Table I, we can rewrite Eqs. (5) and (10) as follows:

$$L\frac{df}{dt} + Rf + \frac{1}{C}\int f\,dt = 0 \qquad (13)$$

B. BLOCK DIAGRAMS

The functional relationship between the variables of a system can conveniently be described in terms of a *block diagram,* in which the parameters of the system are represented by boxes, and the independent and dependent variables by arrows pointing, respectively, into and out of each box. When systems are represented in block diagram form, functions of time are usually transformed into functions of s, the Laplace operator. For example, the elements of the system described by Eq. (13) are described in block form in Fig. 2a. A block diagram of the whole system can be constructed by suitable combination of the elements, as

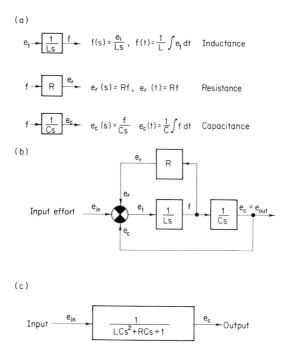

FIG. 2. Block diagram representation of a generalized system: (a) the system components; (b) the synthesized system; (c) the overall system transfer function.

shown in Fig. 2b. A block diagram of this type provides a complete description of the functional relationships between the variables of the system, and the relationship between any two variables can be specified by an equation in terms of s, which is called the *transfer function* of that part of the system. Thus, as illustrated in Fig. 2c, the functional relationship between input and output variables can be described by the overall transfer function of the system. To determine the behavior of the system as a function of time, the transfer function is multiplied by the transformed waveform of the input, and this product is converted to a function of time by application of the inverse Laplace transform. Thus if x is a system input, and y an output, the transfer function $H(s) = y/x(s)$; and $y(t) = \mathcal{L}^{-1}[x(s)H(s)]$, where $x(s) = \mathcal{L}[x'(t)]$. For those readers not familiar with this type of terminology, $\mathcal{L}[x'(t)] = x(s)$ should be read as "the Laplace transform of x' (as a function of time) equals x (as a function of s). Similarly, $\mathcal{L}^{-1}( )$ indicates the inverse Laplace transform of ( ). The purpose of transforming functions of time into functions of s is that it greatly simplifies mathematical manipulations of complex system equations. Indeed, the recent advent of topological methods of systems analysis brings elementary control theory within the range of any biologist with minimal mathematical training. In practice, the transforms of common functions can be looked up in a table thus obviating the necessity of remembering the relevant formula.

The methods employed in the application of control theory to the analysis and description of physiological systems have been detailed extensively by Milhorn (1966) and by Milsum (1966). In addition, the mathematical procedures involved have been outlined in a number of introductory texts (Bayliss, 1966; Machin, 1964; Kalman, 1968; Wilkins, 1966), to which the reader is referred for further information.

III. The Behavioral Element in Homeostasis

Although there have been a number of detailed studies of homeostatic systems in terms of control theory (Crosbie *et al.*, 1961; Grodins, 1959; Pace, 1961; Yates *et al.*, 1968), there has been relatively little such work on behavioral aspects of homeostasis. McFarland (1965a) and Oatley (1967) have touched on some general problems involved in analyzing drinking behavior in terms of control theory; and by virtue of its relative simplicity, the drinking control system is probably the most appropriate homeostatic system in which to start this type of behavioral analysis. Accordingly, the drinking control system will be used here as an example for discussion.

McFarland (1965c) studied the feeding, drinking, and body weight changes in doves in response to a step-change induced by water deprivation. On the basis of the water intake recovery curves alone, it is possible to construct a very simple system for the long-term control of water intake. This system is illustrated in Fig. 3. The system has a single

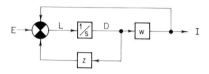

Fig. 3. Simplified version of McFarland's (1965c) system for the control of water intake. E = environmental input, L = rate of water loss, D = water debt, I = rate of water intake, z = the parameter of the conservation feedback loop, and w = that of the ingestion feedback loop; $1/s$ is the Laplace notation for integration.

input E, which represents the combined effect of factors in the environment which influence the rate of water loss. When the animal is maintained with food and water ad libitum at a constant environmental temperature and humidity, the system is in *steady state,* and the output is taken as the mean water intake per unit time. Thus if we use a time unit of 1 day, to avoid complications due to circadian rhythms, the output is measured as the daily water intake. The other variables in the system also have characteristic steady-state values representing the average daily water loss, the average level of water debt, etc. When the animal is deprived of water, the ingestion mechanism no longer functions, and the corresponding parameter *(w)* takes on zero value. In the absence of the ingestive feedback loop, the water debt starts to rise. In addition, the body weight and rate of food intake decline, and by taking these variables into account, it is possible to "dissect" the system into its component subsystems, by applying rules for block diagram or flow graph manipulation (McFarland, 1965a). The system arrived at in this way accounts for many of the features of the interaction of hunger and thirst in this species (McFarland, 1964, 1965b).

In calculating the response of the system to water deprivation, it is necessary to take into account the *initial conditions* of the system. At the instant of change in the value of an input variable, or in the value of a parameter, the state of a system is defined by the values of the state variables. As the state variables are the direct integrals of the rate variables (see above), the number of state variables is the same as the number of integrators in the block diagram of the system. The values of the state variables determine the initial condition of the system at the instant of change, and the number of initial conditions to be taken into account is the same as the order of the system equation. The system described in Fig. 3 is a first-order system, having only one integrator in the block diagram. Thus the state of the system at any time is determined by the value of the water debt. During steady state the value of the debt can be calculated, directly from the block diagram or its equivalent flow graph (see Naslin, 1965; McFarland, 1965a; for this method), or from the system equations (Milsum, 1966). The steady state value of the debt, called D' is given by Eq. 14,

$$D'(s) = \frac{\alpha}{z + w} \tag{14}$$

where α is the value of the system input. The change in water debt during water deprivation is calculated from the block diagram with the ingestive feedback loop removed, and with the above initial condition incorporated. [A simple method for incorporating initial conditions directly into flow graphs is given by Naslin (1965) and by McFarland (1965).] The value of D during water deprivation is given by Eq. (15).

$$D(s) = \frac{D'}{s + z} + \frac{\alpha}{s(s + z)} \tag{15}$$

Substituting in Eq. (14),

$$D(s) = \frac{\alpha}{(s + z)(z + w)} + \frac{\alpha}{s(s + z)}$$

$$D(t) = \mathcal{L}^{-1}[D(s)] = \frac{\alpha}{w + z} e^{-zt} + \frac{\alpha}{z}(1 - e^{-zt}) \tag{16}$$

Equation (16) shows that during water deprivation the water debt rises exponentially from its steady state value. By making similar calculations of the recovery of water intake after deprivation it is possible to calculate the values of the parameters of the system by reference to experimental data. By substituting these values into the appropriate equations it is then possible to make quantitative predictions about the behavior and physiology of the system under various experimental conditions (McFarland, 1965c; McFarland and Wright, 1969). In this way the application of control theory to motivational problems is justified by its predictive power.

It is important to realize that the time-unit employed in the analysis will affect the order of the abstract system. In the above example, a time unit of one day was employed, not only to avoid complications due to circadian rhythms, but also as a general method of simplifying the problem. In employing a large time-unit the order of the system is reduced, because quick-acting transients can be ignored. For example, the delay due to absorption and the transients involved in ingestion cannot be detected in an analysis based on daily measurements. Consequently, the mechanisms responsible for these processes can be regarded as zero-order subsystems and represented by constant parameters. In the case of the parameters designated in Fig. 3, for example, each represents a subsystem, that could be analyzed further in terms of a smaller time unit. A second advantage of this approach is that the system tends to become linearized. Thus drinking, which is really a discontinuous function of time, can be treated as a continuous function for the purposes of the analysis.

This is a general method which could prove to be of particular value in the analysis of motivational systems. From the motivational point of view, the system illustrated in Fig. 3 is a description of the physiological background in which the behavioral element is merely a simple feedback loop. The analysis is only a first step in the investigation of the large and complex system controlling water balance. Further steps, in which a smaller time unit is employed, are directed, either at the detailed dynamics of fluid balance (e.g., Pace, 1961; Reeve and Kulhanek, 1967), or at the more behavioral aspects of the system. As a first step in the latter direction, McFarland and McFarland (1968) analyzed the drinking response of the Barbary dove, employing a time unit of 0.1 hour and paying particular attention to the relative roles of the long-term and short-term satiation mechanisms. A simplified version of their proposed system is illustrated in Fig. 4. In this system the simple ingestive feedback loop of McFarland (1965c) is viewed as a second-order system, in which the ingestion mechanism is represented as a zero-order subsystem (parameter k). Further analysis of this subsystem (McFarland, unpublished), employing a time unit of 1 minute, indicates that the system is essentially nonlinear and involves both saturation and threshold phenomena.

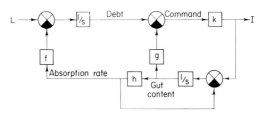

FIG. 4. Simplified version of McFarland and McFarland's (1968) system for the short-term control of the drinking response. L = rate of water loss, and I = rate of water intake, k = the parameter of the ingestion mechanism, g = that of the gut-inhibition mechanism, h = the absorption constant, and f = the hydration coefficient.

Techniques employed in the analysis of the system outlined in Fig. 4 include limitation of the rate of water ingestion, and alteration of the salinity of the drinking water. The rate of water ingestion was measured in subjects trained to obtain 0.1 ml of water for each peck in a Skinner-box. By varying the time-out after each reward, the rate of ingestion could be precisely controlled. The effect of limiting the water intake in this way was calculated in terms of the model (Fig. 5) and the predicted results compared with those recorded when the subjects were run to satiation in the operant situation (Fig. 6). Increasing the salinity of the drinking water was found to be equivalent to increasing the damping ratio of the system, and the effect of the ingestion rate was predicted precisely, as shown in Fig. 7.

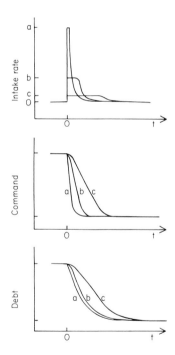

FIG. 5. Step-response of three system-variables at three levels of limitation (a, b, and c) of the rate of water intake. From analog computer simulation of the Fig. 4 system (Mc-Farland and McFarland, 1968).

The philosophy underlying the methods of investigation outlined above is that, rather than try to mimic current physiological or behavioral theories, one should start with the simplest possible hypothesis, and expand this by applying definite rules, in accordance with the results of empirical investigations. Physiological and neurological considerations can be used as a guide in this process, and modifications that make physiological nonsense can be avoided if the investigator always keeps in mind the biological implications of his manipulations. However, there is a danger in relying too heavily on physiological evi-

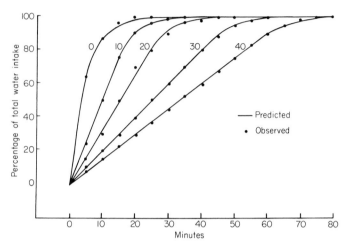

Fig. 6. Experimental limitation of the rate of water intake. Averaged results of 5 subjects at 0, 10, 20, 30, and 40 seconds time-out (from McFarland and McFarland, 1968).

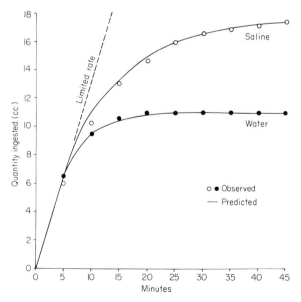

Fig. 7. Form of the integrated drinking response for water and 0.5% saline. Observed results compared with those obtained from analog computer simulation of the Fig. 4 system (McFarland and McFarland, 1968).

dence. Block diagrams represent the equations relating to the behavior, rather than the properties, of particular anatomical structures underlying the behavior, and the danger lies in trying to force an analogy with neurological circuits. The situation is parallel to that found in engineering, where block diagrams are used to describe the behavior of electrical systems, behavior that could be produced by a number of alternative circuits.

IV. HOMEOSTASIS AND MOTIVATION

The influence of homeostatic concepts on motivation theory has been recently reviewed by Cofer and Appley (1964). These authors are primarily concerned with homeostasis as a unifying, all embracing, motivational concept. However, the present discussion is concerned only with those aspects of behavior that are obviously part of homeostatic systems. The problems to be discussed arise logically from consideration of analogies, which result from the use of a common language in the description of physical and biological systems. Control theory provides such a language, and thus acts as a vehicle for these analogies. The relevance of such analogies to behavior comes directly from the inclusion of behavioral mechanisms as integral parts of the homeostatic process. Of particular interest is the concept of motivation and its role in the process of homeostasis.

A. MOTIVATIONAL STATE

The motivational state of an animal sould not be thought of in terms of unitary variables, such as "drive level" (Hinde, 1959). Except in extraordinary circumstances, there are likely to be a number of variables relevant to the state of each motivational system. In the Fig. 4 example, the animal receives information about the state of the blood and about the gut contents. It is reasonable to infer that both these, and other, state variables contribute to the thirst "drive" of the animal.

The familiar, though vague, concept of motivational state can be given rigorous meaning on the basis that a motivational system can be described in terms of control theory. For the present purposes, this assumption is made. As mentioned above, the state of a system can be defined in terms of the state variables of the system, and the number of state variables is determined by the order of the system. In the general case of an nth-order system, the block diagram describing the system would contain n integrators, the outputs of which would be the n state variables. The state of the system is determined by the values of all the state variables, and in principle the system state can be quantitatively described. In practice the state of a system changes continually with time, so that special methods have to be used to portray the state of a system as a function of time. The state of a second-order system can be portrayed on a *state plane diagram,* as illustrated in the following example.

The system portrayed in Fig. 4 is a second-order system, having two state variables D (water debt) and G (gut content). The transient behavior of these variables as a function of time can be determined from the block diagram. In a state plane diagram the state variables are plotted against each other in the manner shown in Fig. 8. The state of the system at any time is indicated by a point on the plane. As the state of the system changes in time, this point traces a *trajectory* on the state plane. By convention, trajectories always move in a clockwise direction. The time at which the system is in a certain state can be marked on the state plane trajectory, so that a complete description of the system state is achieved. In the case of an *n*th order system the state is

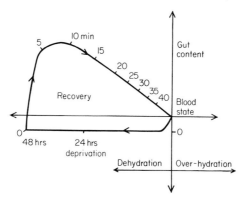

Fig. 8. State plane trajectory for the Fig. 4 system step-response.

represented in a state space of n dimensions. By applying these techniques to motivational systems, drive can be seen as a multidimensional rather than a unitary phenomenon.

B. MOTIVATIONAL ENERGY

The concept of motivational energy has been widely, though loosely, used in the behavioral sciences. Hinde (1960) reviews the energy models of motivation used by Freud (1940), McDougall (1923), and Lorenz (1950), and Tinbergen (1951). Hinde enumerates the confusions and inconsistencies involved in such models, and doubts whether an energy concept is necessary. It is argued here that a rigorous definition of motivational state makes an energy concept logically inevitable. Whether such a concept will prove to be a useful one is another matter.

The first task is to see whether motivational systems can be described in terms of the generalized variables and parameters described above. As an example, consider the system portrayed in Fig. 3. The single state variable of this system, D is readily identified with the generalized displacement h (Table I). Physiologically the water debt can be identified with the state of hydration of the blood, which is displaced from its normal value during water deprivation. As $h = \int f dt + a$ constant, the net and gross rates of water loss can be identified with the generalized flow variable f_L, as shown in Fig. 9. A displacement cannot, of itself, act

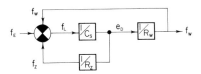

FIG. 9. Generalized representation of the Fig. 3 system.

as a causal agent in a system. The causal agent, a rate variable, must be obtained by measurement of the displacement. Similarly, the state of hydration of the blood does not, of itself, lead to drinking. It is measured by sensory cells which transmit information to the actuating part of the mechanism. Physiologically, it is the osmosity of the blood which is measured. Thus the flow variable f_L should not, strictly, be identified with rates of water loss, but with rates of change of hydration measured as osmotic pressure changes. The significance of this distinction is discussed more fully by McFarland (1965c). As the displacement per se is not an actuating variable, it is omitted from Fig. 9, and the integrating and measuring mechanisms are combined in the generalized capacitance, the output of which is the generalized effort variable e_D, vis:

$$e_D(t) = \frac{1}{C} \int f_L dt, \quad \mathscr{L}\,[e_D(t)] = f_L \cdot \frac{1}{Cs} \tag{17}$$

Thus the measured water debt can be seen as exerting an effort on the ingestion and conservation subsystems. These are represented as the generalized resistances $1/R_w$ and $1/R_z$, having outputs f_w and f_z, respectively. The system portrayed in Fig. 9, in terms of generalized system variables, is consistent with that of Fig. 3. The former is, however, more meaningful psychologically because it makes a distinction between a physiological displacement and a corresponding neurological variable which is identified with the concept of effort.

When the animal is deprived of water the system changes state in accordance with Eq. (16), and in generalized terms (Fig. 9) this equation can be rewritten as follows,

$$e_D(t) = \frac{R_w R_z}{R_w + R_z} \alpha_e^{-t/R_z\,C} + \alpha R_z\,(1 - e^{-t/R_z C}) \tag{18}$$

This equation shows that, during deprivation, the effort level of the system is raised by influx of the environmental flow variable. By analogy with Eq. (11) the energy stored by this process can be calculated from the following formula

$$E_p\ (t)\ =\ \tfrac{1}{2}\,Ce_D{}^2\ =\ \int_0^t P_c dt\ =\ \int_0^t f_L e_D dt, \tag{19}$$

where $P_c = f_L e_D$ is the instantaneous power associated with the capacitance. Thus the level of motivational energy is raised during water deprivation as a result of the continued rate of dehydration f_L. Because the energy storage element is identified with a generalized capacitance, the stored energy is analogous to the *potential energy* of a physical system (Table I). As a general rule, a motivational system will store potential energy when an actuating variable is displaced from its normal (steady state) value.

If the action of the environment were uninhibited during water deprivation (i.e., $f_L = f_E$ in Fig. 9), then all the environmental input would be stored as motivational energy. However, inspection of Fig. 9 shows that $f_L = f_E - f_Z$ during deprivation. The action of the conservation loop thus reduces f_L in proportion to the level of e_D, and the amount of energy stored is consequently reduced by the action of this feedback loop. In generalized terms, energy is dissipated through the resistance R_z in accordance with Eq. (20)

$$E_d\ (t)\ =\ \int_0^t P_r dt\ =\ \int_0^t f_Z e_D dt \tag{20}$$

where $P_r = f_Z e_D$ is the power loss associated with the resistance.

Taking the system as a whole, the energy gained from the environment can be calculated from the equation

$$E_g\ (t)\ =\ \int_0^t f_E e_D dt \tag{21}$$

Substituting $f_E = f_L + f_Z$,

$$E_g\ (t)\ =\ \int_0^t (f_L + f_Z)\ e_D dt\ =\ \int_0^t f_L e_D dt\ +\ \int_0^t f_Z e_D dt \tag{22}$$

Therefore, by Eqs. (19) and (20),

$$E_g\ (t)\ =\ E_p\ (t)\ +\ E_d\ (t) \tag{23}$$

In other words, the sum of the energy stored and the energy dissipated, during deprivation, is equal to the energy gained from the environment.

This result is in agreement with the principle of *conservation of energy*, and an interesting question is raised by the invocation of this principle in discussing motivational systems. What happens to motivational energy that is "dissipated"? In a physical system energy dissipated in a resistor, or by friction, is converted into heat; another form of energy. The environment of a physical system gains the energy dissipated from the system. What is the equivalent process in a motivational system?

This question cannot be answered in its entirety at this stage, as any suggestions made would have to be empirically verified. However, there are two possibilities currently being investigated. The first is that the energy dissipated by one motivational system is gained by another. The energy dissipated during water conservation furnishes a good example of this process. Conservation of water through reduction of evaporative water loss, which is known to occur in some mammals (Schmidt-Nielsen, 1964) and birds (Cade, 1964), results in a reduced heat loss so that, other things being equal, the body temperature tends to rise. This is an important means of water conservation in the camel (Schmidt-Nielsen, 1964), which allows its body temperature to rise considerably during the hot day. The excess heat is stored in the body and is dissipated to the cool environment at night, without use of water. Thus part of the motivational energy dissipated in water conservation is stored in the temperature-regulating system. A similar phenomenon exists in those animals in which reduction of food intake is an important part of water conservation. McFarland (1965c) and McFarland and Wright (1969) have shown that reduction of food intake is the major mechanism of water conservation in the Barbary dove. During water deprivation food intake is considerably reduced, and it is possible to show (McFarland, 1964) that there is a concomitant increase in hunger. Thus part of the motivational energy dissipated in water conservation is stored as potential energy in the feeding system of this species. The details of motivational energy exchange between systems will obviously vary from one species to another, and in order to verify the conservation principle quantitatively, it would be necessary to study a single species intensively.

The second possibility is that information is gained by the animal when motivational energy is dissipated. The rationale for this suggestion comes directly from Brillouin's (1962) negentropy principle of information. In accordance with the second law of thermodynamics, entropy is increased when energy is degraded. In terms of statistical thermodynamics, an increase in entropy represents a transition from an improbable to a more probable, more disordered, state. Entropy can be regarded as a measure of the availability for work of the order within a given system, and is similar in this respect to the availability of information given a number of possibilities. The connection between entropy and information was discovered by Szilard (1929) and rediscovered by Shannon and Weaver (1949). The theme is developed by Brillouin (1962) into a generalization of Carnot's principle: the negentropy principle of information which states that work is required to obtain information and the price paid in increase of entropy will always be larger than the amount of information gained. The sum of negentropy (negative entropy) and information may remain constant in a reversible transformation, but will decrease otherwise.

In applying this principle to the present problem, it is postulated that an animal has to expend motivational energy to obtain information. Moreover, some gain in information is necessary for motivational energy to be dissipated behaviorally. Thus the potential energy increase induced by water deprivation cannot be dissipated behaviorally in the absence of appropriate stimuli, which provide information about the structure of the environment.

C. Ongoing Behavior

From a long-term viewpoint, the behavioral element in a motivational system can be regarded as a zero-order subsystem, as in Fig. 9. In generalized terms, such an element is represented as a resistance through which potential energy is dissipated during ongoing behavior. When the animal drinks in recovery from water deprivation, for example, the energy equation in terms of Fig. 9 will be as follows

$$R_w e_D{}^2 = P_E + \frac{d}{dt}\,(\tfrac{1}{2}Ce_D{}^2) - R_z e_D{}^2 \tag{24}$$

where $R_w e_D{}^2$ is the behavioral power dissipation, P_E is the power input from the environment, $\tfrac{1}{2}Ce_D{}^2$ is the stored potential energy, and $R_z e_D{}^2$ is the power dissipation of the conservation feedback loop.

Each of the dissipative elements in Fig. 9 represents a complex subsystem, the nature of which can be elucidated by means of a more short-term analysis, employing a smaller time-unit, as was done for the system portrayed in Fig. 4. Ongoing behavior is likely to be partly controlled by certain consequences of that behavior, which are monitored by the animal. On this basis we could represent the behavioral elements of Figs. 3, 4, and 9 by a simple feedback mechanism such as that illustrated in Fig. 10a. This type of system would allow the animal to maintain its response rate despite disturbances in the environment. The system is actuated by an *error* variable, which results from comparison of the "desired" and actual consequences of the ongoing behavior. Information about the latter is obtained through a sensory mechanism contained within the *feedback mechanism*. In a Skinner box, for example, a pigeon on a variable-interval reward schedule would have a "desired" rate of pecking corresponding to the *command* in Fig. 10a. Pecks at the response-key would be effective only when the animal pecked with a certain force determined by the experimenter. Effective pecks would result in an auditory stimulus, which could provide the pigeon with information about the consequences of its behavior. In this situation the system illustrated in Fig. 10a would operate so as to maintain the rate of effective pecking at a constant level, because any deviation in the auditorily monitored pecking rate would result in a corresponding change in the error variable, which would actuate the *behavior mechanism* to produce a corresponding change in the pecking force. The experimenter can test the effectiveness of the control system by providing environmental disturbances in the form of changes in the force required to produce an effective response. Chung (1965) found that pigeons in this type of situation can compensate for changes in force requirement up to a

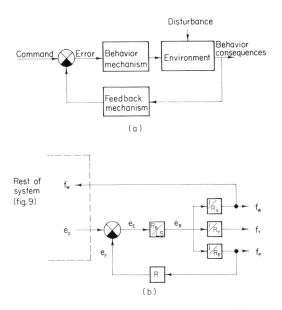

FIG. 10. (a) General description of part of the system controlling appetitive behavior (from McFarland, 1966a). (b) Generalized representation of McFarland's (1966a) system for the control of pecking rate in a Skinner box. See text for further details.

critical value, above which the response rate declines as a linear function of the force requirement. McFarland (1966a) analyzed the behavior observed in these experiments in terms of control theory, but before going on to consider the implications of this analysis, it will be useful to translate the description of the system into generalized terms.

Figure 10b represents McFarland's (1966a) system in generalized terms, seen in the context of an appropriate motivational background (Fig. 9). Thus (translating from food to water reward purely for the purpose of illustration) it is envisaged that when a water deprived pigeon is placed in a Skinner box in which it has learned to obtain appropriate rewards on a variable interval schedule, the link between water debt and command (see Fig. 4) is closed, thus producing a step function in e_D in Fig. 10b. The error variable e_E actuates the behavior mechanism, which is effectively an integrator, written R_B/Ls in generalized terms. The behavior mechanism exerts an effort e_B on the environment, which acts as a resistance. However, different aspects of the environment mediate different behavior consequences, only one of which is of interest here. Thus, in Fig. 10b the environment is depicted as a set of parallel resistances, each with a flow variable output. One of these, f_P, represents the effective pecking rate; while f_W is the rate of water intake and f_Y represents all the other behavior consequences. The rate of effective pecking is transformed into the effort variable e_F by the feedback resistance R_F.

In response to a step input in e_D the rate of effective pecking f_P is given by the following formula,

$$f_P(s) = \frac{\beta R_B}{s(LsR_P + R_F R_B)} \cdot \mathscr{L}^{-1}[f_P(s)] = \beta/R_F \left(1 - e^{-\frac{R_F R_B}{LR_P}t}\right) \quad (25)$$

where β is the height of the step. This expression indicates that a pigeon placed in the Skinner box will peck at a constant rate after an initial warm-up period. The response of the system to sudden changes in the environmental resistance, representing changes in the key-force requirement made by the experimenter, can be calculated, and by finding the appropriate values of the system parameters, quantitative predictions can be made about the response to such disturbances. Figure 11 (top) shows results of this type.

By analogy with Eq. (12) the energy stored by the process described by Eq. (25) can be calculated by the following formula,

$$E_k(t) = \tfrac{1}{2} Lf_P{}^2(t) = \frac{1}{2L} p^2(t) = \frac{1}{2L}\left(\int_0^t e_E dt\right)^2 \quad (26)$$

Thus the level of motivational energy of the behavior mechanism is raised as a result of the step-command, e_D. Because the energy storage element is identified with a generalized inductance, the stored energy is analogous to the *kinetic energy* of a physical system (Table I). This analogy enables us to speak of the *inertial force*, involved in the initia-

tion of a behavior pattern, and of that behavior developing *momentum p.*
Returning to the pigeon pecking response described above. At a critical value of the force-requirement the pigeon is no longer able to maintain the rate of effective pecking, and Chung (1965) found that, above
the critical value, the steady-state pecking rate declined as a linear function of the force requirement. McFarland (1966a) showed that the behavior of the pigeon can be simulated beyond the critical value, by
placing a limit upon the integrator of the behavior mechanism. The
results obtained by this procedure are illustrated in Fig. 11 (bottom).

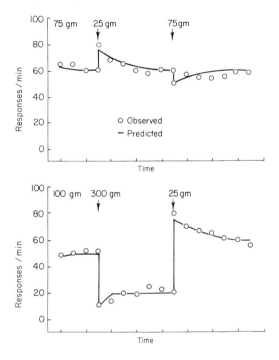

FIG. 11. Changes in response rate which result from changes in force requirement
(shown by numbered arrows), below the critical value (top) and above the critical value
(bottom) (from McFarland, 1966a).

In this case it is the kinetic energy of the behavior mechanism which
is limited, and which is responsible for the decline in response rate as
the environmental resistance increases.

By generalization from this example, we can say that kinetic energy
storage is associated with ongoing behavior, and its immediate consequences. Thus there will be an alteration in kinetic energy storage whenever the consequences of ongoing behavior are disturbed. In terms of
the system outlined in Fig. 10a, reduction of a parameter of the feed-

back mechanism will produce an increase in the error variable, and the animal will appear to "try harder" to maintain the level of behavior consequences. Amsel's (1958) "frustration effect" resulting from non-reward is an example of this phenomenon.

The main problem posed by the concept of motivational energy is the behavioral meaning of energy dissipation. Energy is dissipated through a behavioral resistance, when the energy level of the system is reduced. The problem is aggravated by the fact that the so-called behavioral resistances are in fact subsystems containing smaller energy storage elements, the behavior of which demands more detailed analysis. However, in general we can say that, as in physical systems the kinetic energy level of a subsystem is increased temporarily as potential energy is dissipated through it. However, as in physical systems, the usefulness of such concepts can be evaluated only through rigorous quantitative experimental analysis of the behavior of the system. Although some steps have been taken in this direction, validation of the proposals outlined in this paper must await further empirical verification.

V. Conclusions

The main arguments put forward in this paper are as follows: (1) The existance of dynamic analogies between physical systems enables us to define generalized variables and parameters, in terms of which any dynamic system may be described. (2) Description of motivational systems in generalized terms can be achieved through the application of control theory to the analysis of behavior as a function of time. (3) By delineating the important state variables in a motivational system, the concept of motivational state can be given rigorous meaning. (4) Consideration of the changes of state involved in behavioral transients leads inevitably to concepts of work, power, and energy, analogous to those in the physical sciences.

In view of the lack of detailed experimental evidence it may be argued that speculation about motivational energy is premature. In reply, there are three main points to be made: First, if it is agreed that motivational systems can be described in terms of control theory, then the concept of motivational energy, as defined mathematically, follows logically from the representation of such systems in generalized terms. If it is denied that control theory is relevant to behavior, then we have to postulate a fundamental discontinuity in the nature of homeostatic mechanisms. This question can only be answered empirically, and the present evidence suggests that it is possible to describe motivational systems, at least at the gross level, in terms of the behavior of the relevant variables as a function of time. However, the mathematical validity of the motivational energy concept does not mean that the concept is a useful one. This is very much an open question and brings us to the second point. By introducing a motivational energy concept at this stage, the way is opened for a *post hoc* identification of the energy exchanges involved. For example, the vacillations observed in a conflict situation immediately suggest a reciprocal exchange of potential and kinetic energy, and even if it is not possible to analyze the system in a conventional manner, it may be possible to work backward from this supposition. Finally, the energy concept has considerable intuitive appeal, as evidenced by its frequent use in the past. Concepts such as energy storage, and behavioral inertia and momentum, which make behavioral sense intuitively, need not be used vaguely, and can be given a precise meaning in terms of the observed and postulated variables involved.

References

Amsel, A. 1958. The role of frustrative non-reward in noncontinuous reward situations. *Psychol. Bull.* **161**, 374–386.

Bayliss, L. F. 1966. "Living Control Systems." English Univ. Press, London.

Bendat, J. S., and Piersol, A. G. 1966. "Measurement and Analysis of Random Data." Wiley, New York.

Brillouin, L. 1962. "Science and Information Theory," 2nd Ed. Academic Press, New York.

Cade, T. J. 1964. Water and salt balance in granivorous birds. *In* "Thirst—Proceedings of the 1st International Symposium on Thirst in the Regulation of Body Water." (M. Wayner, ed.), pp. 237–256. Macmillan (Pergamon), New York.

Chapman, K. M., and Smith, R. S. 1963. A linear transfer function underlying impulse frequency modulation in a cockroach mechanoreceptor. *Nature (London)* **197**, 699.

Chung, S. H. 1965. Effects of effort on response rate. *J. Exp. Anal. Behav.* **8**, 1–7.

Cofer, C. N., and Appley, M. H. 1964. "Motivation: Theory and Research." Wiley, New York.

Crosbie, E. J., Hardy, J. D., and Fessenden, E. 1961. Electrical analog simulation of temperature regulation in man. *IRE Trans. Bio-Med. Electron.* **8**(4).

Freud, S. 1940. "An Outline of Psychoanalysis." Hogarth Press, London.

Grodins, F. S. 1959. Integrative cardiovascular physiology: A mathematical synthesis of cardiac and blood vessel hemodynamics. *Quart. Rev. Biol.* **34**, 93–116.

Hinde, R. A. 1959. Unitary drives. *Anim. Behav.* **7**, 130–141.

Hinde, R. A. 1960. Energy models of motivation. *Symp. Soc. Exp. Biol.* **14**, 199–213.

Hinde, R. A. 1966. "Animal Behaviour. A Synthesis of Ethology and Comparative Psychology." McGraw-Hill, London.

Kalman, R. E. 1968. New developments in systems theory relevant to biology. *In* "Systems Theory and Biology" (M. D. Mesarovic, ed.), pp. 222–232. Springer, Berlin.

Korn, G. A. 1966. "Random-Process Simulation and Measurements." McGraw-Hill, New York.

Lorenz, K. 1950. The comparative method of studying innate behaviour patterns. *Symp. Soc. Exp. Biol.* **14**, 221–268.

McDougall, W. 1923. "An Outline of Psychology." Methuen, London.

McFarland, D. J. 1964. Interaction of hunger and thirst in the Barbary dove. *J. Comp. Physiol. Psychol.* **58**, 174–179.

McFarland, D. J. 1965a. Flow graph representation of motivational systems. *Brit. J. Math. Statist. Psychol.* **18**, 25–43.

McFarland, D. J. 1965b. The effect of hunger on thirst motivated behaviour in the Barbary dove. *Anim. Behav.* **13**, 286–292.

McFarland, D. J. 1965c. Control theory applied to the control of drinking in the Barbary dove. *Anim. Behav.* **13**, 478–492.

McFarland, D. J. 1966a. A servoanalysis of some effects of effort on response rate. *Brit. J. Math. Statist. Psychol.* **19**, 1–13.

McFarland, D. J. 1966b. On the causal and functional significance of displacement activities. *Z. Tierpsychol.* **23**, 217–235.

McFarland, D. J. 1967. Phase relationships between feeding and drinking in the Barbary dove. *J. Comp. Physiol. Psychol.* **63**, 208–213.

McFarland, D. J., and McFarland, F. J. 1968. Dynamic analysis of an avian drinking response. *Med. Biol. Eng.* **6**, 659–668.

McFarland, D. J., and Wright, P. 1969. Water conservation by inhibition of food intake. *Physiol. Behav.* **4**, 95–99.

Machin, K. E. 1964. Feedback theory and its application to biological systems. *Symp. Soc. Exp. Biol.* **18**, 421–445.

Milhorn, H. T., Jr. 1966. "The Application of Control Theory to Physiological Systems." Saunders, Philadelphia, Pennsylvania.

Milsum, J. H. 1966. "Biological Control Systems Analysis." McGraw-Hill, New York.

Naslin, P. 1965. "The Dynamics of Linear and Non-Linear Systems." Blackie, London.

Oatley, K. 1967. A control model for the physiological basis of thirst. *Med. Biol. Eng.* **5**, 225–237.

Pace, W. H. 1961. An analogue computer model for the study of water and electrolyte flows in the extracellular and intracellular fluids. *IRE Trans. Bio-Med. Electron.* **8**(1).

Pringle, J. W. S., and Wilson, V. J. 1952. The response of a sense organ to a harmonic stimulus. *J. Exp. Biol.* **29**, 220–234.

Reeve, E. B., and Kulhanek, L. 1967. Regulations of body water content: A preliminary analysis. *In* "Physical Basis of Circulatory Transport: Regulation and Exchange." (E. B. Reeve and A. C. Guyton, eds.). Saunders, Philadelphia, Pennsylvania.

Schmidt-Nielsen, K. 1964. "Desert Animals. Physiological Problems of Heat and Water." Oxford Univ. Press (Clarendon) London and New York.

Shannon, C. E., and Weaver, W. 1949. "The Mathematical Theory of Communication." Univ. of Illinois Press, Urbana, Illinois.

Stark, L., Vossius, G., and Young, L. R. 1962. Predictive control of eye tracking movements. *IRE Trans. Hum. Factors Electron.* **3**(2), 52–57.

Stark, L. 1959. Stability, oscillations, and noise in the human pupil servomechanism. *Proc. IRE* **47**, 1925–1936.

Stark, L., and Young, L. R. 1964. Defining biological feedback control systems. *Ann. N.Y. Acad. Sci.* **117**, 426–442.

Szilard, L. 1929. Uber die Entropieverminderung in einem thermodynamishen System bei Eingriffen intelligenter Wesen. *Z. Phys.* **53**, 840–856.

Thorson, J. 1966. Small-signal analysis of a visual reflex in the locust. *Kybernetik* **2**, 41–53, 53–66.

Tinbergen, N. 1951. "The Study of Instinct." Oxford Univ. Press, London and New York.

Wilkins, B. R. 1966. Basic mathematics of control. *In* "Regulation and Control in Living Systems" (H. Kalmus, ed.). Wiley, New York.

Yates, F. E., Brennan, R. D., Urquhart, J., Dallman, M. F., Li, C. C., and Halpern, W. 1968. A continuous system model of adrenocortical function. *In* "Systems Theory and Biology" (M. D. Mesarovic, ed.). Springer, Berlin.

Young, L. R., Green, D. M., Elking, J. L., and Kelly, J. A. 1963. Adaptive characteristics of manual tracking. *4th Nat. Symp. Hum. Factors Electron., Washington, D.C.*

Individual Recognition of Voice in the Social Behavior of Birds

C. G. Beer

RUTGERS UNIVERSITY, NEWARK, NEW JERSEY

I. INTRODUCTION

In a paper that deserves to be better known, Wallace Craig discussed the functions served by vocalizations in the social behavior of pigeons (Craig, 1908). He concluded that the organization of social behavior in pigeons is more flexible than regulation by simple instinctive mechanisms would allow, and that much of this flexibility resides in the complex and extremely varied use of voice by these birds. He found the uses of vocalizations to be so varied and complexly interrelated that he could give only what he considered to be an incomplete and oversimplified list of them.

In spite of the profound influence Craig had on European ethology, however, it has been only in recent years that ethological research has begun seriously to take up the challenge presented by the kind of observations in this paper on pigeon calls. Indeed it could be argued that what classical ethology took from Craig's writing directed attention away from the kind of complexity he was so much aware of in the social behavior of birds. His famous paper on "appetites and aversions" (Craig, 1918) provided the concepts of appetitive behavior and consummatory act

27

that were incorporated in central propositions of the instinct theories of Lorenz (e.g., 1937, 1950) and Tinbergen (e.g., 1951). But the concern of these theories with stereotyped and species-specific characteristics of behavior, with social communication conceived of as a system of displays and responses innately tuned to one another, did little to elucidate or encourage investigation of such matters as the nature of the relationships between individuals in social groups, the development and maintenance of such relationships, and their contribution to social organization.

In his list of the uses of pigeon "song," Craig included the following: "To proclaim: (a) the bird's species; (b) the bird's sex; (c) the bird's individual identity; (d) the bird's rights" (Craig, 1908, p. 99). Species recognition, the recognition of sex, and the expression of motivation are matters to which ethological theory and research directed considerable productive attention. But individual recognition, being outside the range of phenomena regarded as instinctive, attracted little notice until the influence of instinct theory on ethological thinking began to decline.

Individual recognition in the social behavior of birds was not entirely ignored. Field observations suggesting its existence were reported (e.g., Goethe, 1937; Tinbergen, 1953), and it was at least implied by work on dominance hierarchies and territorial behavior. But as recently as 1963, Busnel had to write ". . . as yet there are no precise experiments . . ."; and Bremond (1963) could find only field observations to include in a short review of the subject. It was not until a few years ago that questions about the characteristics upon which individual recognition is based, the nature and timing of its development, and its functional significance, began to be subjects for experimental enquiry and quantitative analysis. The results of this enquiry suggest that individual recognition is widespread in the social behavior of birds, and that its importance in the statics and dynamics of avian social organization is much greater than was allowed by accounts of social behavior viewed as the product of instinctive mechanisms.

But theoretical orientation has almost certainly not been the only factor responsible for the relative neglect of individual recognition in the social behavior of birds. So far the experimental and analytical work on individual recognition in birds has been concerned almost exclusively with the auditory modality. The fact that birds are perhaps the most spectacularly vocal creatures in the animal kingdom only partly explains this bias. Equipment for the recording, reproduction, and analysis of sound is now available that combines high fidelity and precision with ease of operation and portability. Consequently the measurement and description of a stimulus, and the reproduction of the stimulus for experimental purposes, can be more readily accomplished for sound stimuli than for stimuli in other modalities. Perhaps we have here an instance of how technology can influence the direction of research. And it might be argued that at least part of the reason for the late start of investigation into individual recognition in bird behavior was that the research had to wait on the invention of tools for carrying it out.

These thoughts can be applied to the work on imprinting. The classical studies on imprinting were set in the context of the question: How does an animal acquire the ability to recognize a member of its own species? (e.g., Lorenz, 1935) even though what has been abstracted as imprinting may very well be part of a process which, in nature, leads to individual attachments. This orientation may have been as much a consequence as a cause of the fact that the differences between stimulus objects used in these studies were on the scale of species differences rather than individual differences. Most of the experiments on imprinting have been concerned with visual stimuli. Individual characteristics in the appearances of animals are difficult for us to discern, describe, and duplicate. Perhaps preoccupation with visual stimuli is a reason why the relationship of imprinting to the formation of individual attachments has received little attention. One of the few studies that has shown individual attachment to be the outcome of imprinting dealt with olfactory stimuli in a mammal (Klopfer and Gamble, 1966), not visual stimuli in a bird, and is thus an exception that proves the rule. However, even when auditory stimuli have been involved in imprinting studies of birds, the manipulations have been between species differences rather than individual differences (e.g., Gottlieb, 1965). The studies of individual recognition of voices in birds have, for the most part, come from research orientations other than the pursuit of imprinting. Some of these studies may set the subject of imprinting in a broader functional and developmental perspective than it has been customary to consider so far.

However, before the importance of individual recognition in the social behavior of birds can be assessed, we shall need to know more about it than we do at present. Our knowledge is as yet too fragmentary for the writing of a balanced review. In this essay I shall survey the various ways in which individual recognition of voice in birds has been studied and in so doing attempt to convey the breadth of this field of research. Then I shall describe a study of my own in sufficient detail to illustrate how rich a complex of facts and problems can be discovered by research on vocal recognition in a single species. Finally I shall discuss some of the findings in the light of comparative, functional, and developmental questions.

II. Sources of Evidence

Three approaches have been employed, either singly or in combination, in the investigation of vocal recognition in birds: field observation, sound analysis, experiments to compare a bird's reactions to the sounds of different individuals.

A. Field Observation

Observations of wild birds living in nature, and of captive birds living in seminatural conditions, was, for a long time, virtually the only source of evidence of individual recognition of voice in birds (see Bremond, 1963, pp. 720–721). It still, as a rule, provides the point of departure for the other two methods of investigation, directing attention to the kinds of vocalizations that might convey individual identity and to the kinds of behavioral interactions into which individual recognition of voice might enter. If an ornithologist can learn to identify individual birds by ear, one supposes that the birds can too. If a bird is observed to react differently to the same kinds of calls uttered by different individuals, even when the birds are out of sight of one another, one has at least circumstantial evidence of recognition based on individual characteristics of voice. The ornithological literature contains many instances of both these kinds of observation.

According to Thorpe (1961, p. 100): ". . . differences in the Chaffinch song are individually characteristic . . . and the experienced ear can . . . recognize a given bird with almost complete certainty, as by a signature tune." Similar observations have been made of other song birds, e.g., American wood warblers (Kendeigh, 1945; Stewart, 1953; Borror, 1961), the song sparrow, *Melospiza melodia* (Nice, 1943), the rufous-sided towhee, *Papilo erythropthalmus* (Armstrong, 1963, p. 92), and the magpie lark, *Grallina cyanolenca* (Robinson, 1946, 1947). Individual characteristics in the calls of some nonpasserine species are also sufficiently distinct for individual recognition by the human ear. Beebe (1918–1922; also cited in Armstrong, 1963, p. 76) could identify individual argus pheasants *(Argusianus argus)* in this way. In a feral population of domestic fowl *(Gallus)* McBride (personal communication) was able to identify individual cocks by their crows as well as by their locations.

While observations like these leave no doubt of the existence of individual differences in the vocal behavior of certain species, they allow no very precise statements about the nature of these differences; nor do they indicate to which of the differences, if any, the birds are sensitive. The study of these questions requires the sorts of analytical and experimental approach that I shall come to directly. But such study has been encouraged by field observations of the ways in which sounds can affect the behavior of birds.

It was known before the invention of the tape recorder that in many species the birds are sensitive to fine detail in auditory stimuli and can appreciate fine differences between the sounds they hear. Great crested Grebes *(Podiceps cristatus)* are reputed to be capable of distinguishing the sounds of individual motor boats on a lake (Buxton, 1946). The learning of sounds manifested in the various forms of vocal mimicry in birds implies, in many cases, a fine ear for nuances of pitch, timbre, rhythm, and sound pattern (for discussions of vocal mimicry in birds see Armstrong, 1963; Thorpe, 1967). In some species of song birds (e.g., the chaffinch, *Fringilla coelebs*) in which an individual possesses several songs or variants of songs in its repertoire, a bird replying to the singing of a rival will usually do so with that song or variant in its repertoire that matches most closely the utterance of the rival (Hinde, 1958; Thorpe, 1958).

More direct evidence of recognition based on individual characteristics of voice is provided by field observations of birds apparently responding selectively to the songs or calls of particular individuals. According to reports, individual recognition by ear can be found between the members of a mated pair, for example, the ring dove, *Streptopelia risoria* (Craig, 1908); the turnstone, *Arenaria interpres* (Bergman, 1946); herring gull, *Larus argentatus* (Tinbergen, 1953), and the bullfinch, *Pyrrhula pyrrhula* (Nicolai, 1956). It has been observed between rival territorial males in, for example, the pied flycatcher, *Muscicapa hypoleuca* (Curio, 1959); and the cactus wren, *Campylorhyncus brunneicapillus* (Anderson and Anderson, 1957). And it has been observed between parents and their young, either recognition of the young by their parents, as in the herring gull (Goethe, 1937), sooty terns, *Sterna fuscata* (Watson and Lashley, 1915), and the jackdaw, *Corvus monedula* (Lorenz, 1931, 1938); or recognition of parents by their young as in the herring gull (Goethe, 1937), jackdaw (Lorenz, 1931, 1938), domestic fowl (Brückner, 1933; Collias, 1952) and European blackbird, *Turdus merula* (Messmer and Messmer, 1956; Thielke-Poltz and Thielke, 1960). Further references will be found in the books by Thorpe (1961) and Armstrong (1963).

However, many of the reports of field observations leave open the question to what extent selective responsiveness to particular individuals is exclusively a matter of recognition of individual vocal characteristics. The possibility that visual characteristics might at least support the vocal characteristics cannot be ruled out in situations in which the birds can see as well as hear one another. That visual characteristics can substitute for vocal characteristics in the conveying of identity has been observed in the herring gull (Tinbergen, 1953). In the ruff *(Philomachus pugnax)* individual differences in visible characteristics may provide the only means by which females can identify individual males on the lek, since the males are not very vocal (Armstrong, 1963, p. 77).

But even in situations in which the birds are not visible to one another, differential responsiveness to the calls or songs of individuals may not be due to recognition of the voices, or this alone. The location from which a sound comes may be at least as important as the nature of the sound itself, particularly in territorial species. For instance, if each territory-holding bird in an area delivers its songs from a set of specific locations — its "singing posts" — the song of a nonresident will be distinguishable from those of residents if it occurs at a locality from which song does not customarily come, or it will elicit hostility from the particular bird whose locality has been trespassed.

On the other hand, field observations can lead to the conclusion that individual recognition is lacking in situations in which, in fact, it exists. Until recently it was believed that, in those species of penguins in which the young aggregate into large groups with only a few adults in attendance, the young of such a group are fed indiscriminately by the adults, the relative hunger and hence activity of the chicks, rather than family relationships determining who was fed by whom on any particular occasion. It has now become apparent from closer observation and experimental studies that even in these "creche" situations the rule is that young are fed only by their own parents (Richdale, 1957; Sladen, 1958; Stonehouse, 1960; Penney, 1962; Thompson and Emlen, 1969). Ideas about parental care in Sandwich terns *(Sterna sandvicensis)* have been similarly revised (Tinbergen, 1953).

Another problem that field observation may be insufficient to solve, or may suggest the wrong answer to, is that although patterns of interaction among individuals of a social group may be such as to entail individual recognition, it may not be clear who recognizes whom. In his book on the herring gull, Tinbergen (1953) reproduced some field notes on the effects of interchanging young of different nests, from which he concluded that the parent gulls ". . . learn to know their own young . . ." by about the fifth day post-hatching. But his observations would be equally consistent with the possibility that the older young were discriminating between their own parents and the strange adults, or were reacting to being placed in a strange locality, with the consequence that their behavior rather than any individual characteristics marked them as foreign to the strange adults and thus stimulated the hostility upon which the claim that the parents recognized their own chicks was based.

Field observation then, while leaving little doubt that individual recognition involving voice occurs in the social behavior of birds of many species, raises many questions about it that can be decided only by other approaches.

B. ANALYSIS OF VOCALIZATIONS

The analysis of bird vocalizations has been revolutionized by the invention of machines that can transform recorded sounds into visual forms. Such "pictures" of sounds enable one to describe and measure the sounds in more detail and with more precision than reliance on hearing can even remotely approach; and they can bring to light features of sound that the human ear is incapable of registering but to which the avian ear may well be sensitive (Greenwalt, 1968; Marler, 1969).

Nevertheless, the instruments that are used to make these pictures also have their limitations. The cathode-ray oscilloscope can give a graphic representation of the amplitude of sound intensity as a function of time but ignores all other parameters of sound. The conventional sound spectrogram is a more complete picture, plotting frequency and, in a less precise way, intensity as functions of time. But one has to choose between alternative renderings. On the machines in most general use (Kay Electric Sonagraph 661A) there are two bandpass filter settings: by selecting the wide bandpass filter (bandwidth 300 Hz) one obtains a relatively sharp rendering of temporal pattern (resolution about 3 msec) but not a very detailed picture of frequency structure and "shape"; by selecting the narrow bandpass filter (bandwidth 45 Hz) one gets a relatively detailed rendering of frequency structure and shape but has to accept some smudging in the temporal dimension (resolution about 25 msec) [for fuller accounts of sound analysis by means of the sonagraph, see Borror and Reese (1953), Cherry (1966), and Marler (1969)]. The frequency range of the machine is 85–8000 Hz. The presence of higher frequencies in a recorded sound can be displayed in a sonagram if the sound is played into the sonagraph at a speed slower than the recording speed, but then one sacrifices detail in the lower frequency range. More recent models of the sonagraph make available a wider selection of frequency ranges, filter settings and temporal settings, and other features. But there is still no way of portraying a sound in a single picture that is complete in all respects. The investigator has to decide what features of a sound are most likely to carry the information he is searching for and choose his method of graphic analysis accordingly.[1] Similar points could be made about the choice of recording microphone, tape recorder, and so on (see Andrieu, 1963).

Where there is reason to believe that the information is coded in the form of amplitude modulation, or on-off patterning, as appears to be the case in acoustic communication in insects (e.g., Dumortier, 1963), the osillogram is an appropriate choice. Tschanz (1965, 1968) made oscillograms of "Lockrufe" of guillemots *(Uria aalbe)* and found that the temporal patterning of a call — the durations of the pulses of sound and their order of succession — is characteristic of an individual. But the sonagraph can also be used to obtain a display of amplitude as a function of time and this has been done for the landing calls of gannets *(Sula bassana)* (White and White, 1970; White *et al.*, 1970). The "amplitude envelopes" of calls given by male gannets on separate occasions were compared using linear correlation. It was found that the correlation between calls of the same individual was higher than between those of different individuals, and that the first part of a call contained sufficient information to identify the caller.

Most analyses of bird vocalizations, however, have focused on frequency/time patterns, and it has been for the display of these that the sonagraph has had its fullest employment. Perhaps because precise temporal specification has generally been regarded as more important for analysis than fine grain in the frequency spectra, wide-band filter settings have been used more often than narrow-band settings in the production of sonagrams of bird calls and songs. There is reason to believe that the temporal resolution of the avian ear is remarkably fine; auditory reaction times may be as much as ten times faster in birds than thay are in men (Pumphrey, 1961; Schwartzkopf, 1962; Greenwalt, 1968; Marler, 1969). Where there is suspicion that temporal features of vocalizations might be involved in individual recognition, therefore, it is appropriate that the finest temporal resolution available be used for analysis. Such analysis of antiphonal singing in certain species has shown that in the duets sung by the members of a mated pair, although other details may vary, the time interval between the beginning of the contribution of the first bird and the beginning of the contribution of the second is remarkably constant e.g. in eight consecutive duets sung by a pair of *Laniarius erythrogaster,* the mean "response time" of the

[1] A machine has recently been described (Hjorth, 1970) that produces a frequency/time display ("melogram") with much less detail than a sonagram. The comment above applies to this machine also.

second bird was 144 msec with a standard deviation of only 12.6 msec (Thorpe, 1963). In a closely related species, the gonelek *(L. barbarus),* Grimes (1965, 1966) has found comparable constancy in duet response times and also distinctly different response times between different pairs. These findings have led to the suggestion that in these species the second bird identifies itself to the first by the timing of its answering call, and that in this way the birds of a pair can keep in auditory communication with one another and thus maintain the pair bond in a habitat in which the foliage is so dense that the birds are not visible to one another much of the time. However, Hooker and Hooker (1969) have argued that response time is probably not the only feature serving the function of individual recognition in the duet singing of these species.

Individual characteristics in the temporal patterning of vocalizations is not confined to antiphonal singing. For example N. S. Thompson (1969) has found that in the American common crow (*Corvus brachyrhynchos*) the durations of caws and of the intervals between successive caws and other temporally specifiable features are individually distinct. This discovery supports a suggestion that cawing enables the members of a foraging flock to identify one another individually and thus to control their dispersion.

Important as temporal patterning appears to be as a means of conveying individuality in many species, however, individual characteristics of pitch or patterns of pitch change often strike our ears even more forcibly. The use of the narrow-band filter setting of the sonagraph has led to relatively precise specification of frequency characteristics of vocalizations in some species. Here too individual differences have been found. For example, such differences have been found in the song of the white-throated sparrow (*Zonotrichia albocolis*) (Borror and Gunn, 1965; Falls, 1963, 1969), and in the antiphonal calls of the eastern whipbird (*Psophodes olivaceus*) (Watson, 1969).

An alternative method of analyzing frequency spectra of bird vocalizations has been devised and used by Marler and Isaac (1960a). The standard sonagraph has a feature that enables one to make a section of amplitude/frequency at selected instants of a recording, but these have to be at least 40 msec apart from one another. Marler and Isaac added a modification to the machine that allowed them to make amplitude/ frequency sections at intervals as small as 2.5 msec. Using this device and the narrow bandpass setting they were able to produce sonagrams in which the distribution of sound energy in the frequency range was displayed in detail (compared to the standard frequency/time sonagram in which it is represented by the darkness of shading of the trace), albeit on a relatively coarse time scale. With this technique and the more conventional sonagrams they were able to show that the songs of chipping sparrows (*Spizella passerina*) have individual characteristics that could be used in individual recognition.

Other investigators have made both wide and narrow band sonagrams of the same calls, and so have been able to obtain relatively precise measurements of both temporal and frequency characteristics (e.g., Borror and Gunn, 1965; Hutchison *et al.*, 1968). In the study by Hutchison *et al.* (see also Thorpe, 1968) the "fish calls" of Sandwich terns were recorded as the birds were flying in from the sea to feed their young. Two successive calls of forty birds were analyzed. From the two types of sonagrams seven different characteristics were measured. Each showed high positive correlation between the calls of the same individual ($r > 0.80$). On the assumption that characteristics with the greatest variability in proportion to their average values are the most readily discriminated, the investigators argued, from comparisons of the ratios of standard deviation/mean for the seven measures, that three of the characteristics (duration and number of vertical bars in the third segment, and lowest frequency of the fundamental of the second segment) probably serve the function of conveying identity.

Even with sonagrams, however, it is often difficult to describe sounds in purely quantitative terms. The shapes of the components of vocalizations may be so complex and various as to defy measurement except on a nominal scale in which assignments are made on the basis of visual judgments of similarity and difference of form. Even where more precise measurement is possible, immediate impressions of "gestalt" are sometimes all that one needs to sort a sample of sonagrams into distinct groups. In this way Evans (1970a, personal communication) arrived at the conclusion that in black-billed gulls (*Larus bulleri*) and ringbilled gulls (*Larus delawarensis*) the "mew calls" of adults are individually distinct. In both cases the similarities between sonagrams of calls of the same bird and the differences between sonagrams of calls of different birds were such that visual inspection alone was sufficient for naive viewers to sort the sonagrams into groups which corresponded to the individual birds.

Such analyses as these I have briefly described have thus provided evidence of individual differences in vocalizations in a number of bird species ranging from penguins to song birds. In some species the individual characteristics remain remarkably stable from year to year (Borrow, 1960; Thorpe, 1958, 1961). Sonographic analysis has also enabled investigators to make explicit in quite precise ways the particular respects in which the vocalizations of individuals are distinct, and to distinguish these from characteristics that are species specific. For example, in a study of song variation in the brown towhee *(Papilio fuscus)* Marler and Isaac (1960b) found that whereas duration and temporal pattern of song are almost invariant from one individual to another, and might therefore serve for species recognition, syllable characteristics such as frequency pattern vary markedly between individuals and so provide a possible means of individual recognition. From the same sort of evidence, it has been suggested that in the songs of the *Hylocichla* thrushes species identification is conveyed by temporal pattern, duration of syllables, and relative changes of pitch, while individual identity is conveyed by detailed structure (Stein, 1956).

The existence of individual characteristics is a necessary condition for individual recognition of voice in birds, but by itself it is not sufficient to prove that such recognition occurs in any particular case. The next step is to put the question to the bird in the form of an experiment.

C. Experimental Studies

Apart from some field experiments involving the placing of young with adults other than their parents (e.g., Lashley, 1915; Goethe, 1937), 1954; Tinbergen, 1953; Rittinghaus, 1953; Dircksen, 1932; Collias, 1952; Davies and Carrick, 1962) experimental approaches to the question of whether birds can recognize one another individually by ear have used playback of recorded sound. This technique enables the investigator to test the effects of the sound itself in isolation from other factors that might normally be associated with it, such as specific visual stimuli and specific locations, which might at least contribute to selective responsiveness.

Two methods of playback presentation have been employed: successive presentations of recordings of different individuals, and simultaneous presentations of recordings of different individuals. Rarely have both forms of presentation been included in the same study. In one instance I know of (Evans, 1970b), the results of the two forms of presentation were different: ring-billed gull chicks failed to distinguish the "mew calls" of their parents from those of other gulls in successive discrimination playback tests, but did show recognition of the voices of their parents in the simultaneous tests. This difference in the results of the two kinds of tests will come as no surprise to anyone familiar with the comparisons of performance on successive and simultaneous discrimination tasks in the learning theory literature (e.g., Kimble, 1961).

In general it appears that simultaneous discrimination is more easily achieved than successive discrimination, and therefore that simultaneous presentation provides a more sensitive test of ability to discriminate two stimuli than does successive presentation.

However, there are also instances where the two kinds of test give conflicting results if they are regarded as measuring the same thing. Stellar (personal communication), in studies of sweetness preference in the feeding of human subjects, has found a situation in which, if preference is judged by which of two foods differing in sweetness is consumed in greatest quantity in successive tests, the less sweet of the two is preferred; whereas if preference is judged by choice between the two foods offered together, the sweeter is preferred. This sort of result suggests that one should be cautious about the conclusions on individual recognition that one draws from only one type of discrimination test. The possibility should be kept in mind that a bird may respond differently in a situation in which both familiar and unfamiliar individuals are present, from the way in which it responds when only one or the other are present, and that there may be sound functional reasons why this should be the case. For example, Evans (1970b) has pointed out that if a gull chick has lost its parents its only hope of survival lies in the chance that it can get itself accepted and fostered by other adults, and for this to occur it will not be in the chick's interests for it to persist in the discrimination against other gulls that appears to be conducive to social harmony when families are intact.

Playback experiments to test whether birds can recognize one another individually by ear have been carried out on adult birds and young birds. Since the problems presented by interactions between adults are different from those presented by interactions between adults and young I shall treat them separately.

1. Experiments on Adults

Although playback of sounds has been widely used as a technique for stimulating birds to call or sing, it has most often been for purposes other than the investigation of individual recognition, such as to find out whether birds from different populations, between which the degree of taxonomic divergence is unsettled, can discriminate between vocalizations of the different populations [e.g., Thielke (1962) on the New World and Old World creepers (*Certhia*); Dilger (1956) on the *Hylocichla* and *Catharus* thrushes; Lanyon (1963, 1967) on the *Myiarchus* flycatchers; other references in Lanyon (1969)], and to discover what features of a vocal pattern are necessary or sufficient for it to evoke response [e.g., Abs (1963) on the European nightjar (*Caprimulgus europaeus*); Bremond (1967) on the European robin *Erithacus rubecula*]. However some of these studies have also involved or led to comparison of the effects of playback of songs or calls of different individuals of a bird's own group. As a result, we now have evidence of individual recognition of voice both in the context of territorial advertisement and defense and in the maintenance of pair bonds.

Weeden and Falls (1959) tested territory-holding male ovenbirds (*Seiurus aurocapillus*) with successive playback of recordings of their own songs, the songs of males holding neighboring territories, and the songs of males holding territories far enough away to be virtually out of earshot. The tested birds sang in reply to the songs played to them. They responded most strongly to the songs of the distant males; least strongly to the songs of their neighbors; and their responses to the recordings of their own song were intermediate.

From this evidence of differential responsiveness, however, the conclusion does not necessarily follow that an ovenbird male can recognize a particular individual by its song. There is no need to assume more than that the birds distinguish familiar songs as a class from unfamiliar songs. The intermediate position of response to playback of a bird's own voice can be viewed as consequent on the fact that in this case the song was familiar in some respects and not in others, to the bird. In any case it could be argued that differential responsiveness to neighbors and strangers is all that is required for functionally efficient territorial defense.

However a more recent study, on the white-throated sparrow, has provided evidence of individual recognition that is more specific. Brooks (cited in Falls, 1969) has found that, as in the ovenbird, male white-throated sparrows distinguish between the songs of neighbors and the songs of strangers, but also that response to playback of the song of a neighbor varies with the location of the loudspeaker. If the song is played from the direction of the neighbor's territory, response is weak; if the song is played from the opposite direction, response is as strong as to the song of a stranger. This result makes two points: the birds must be able to distinguish between the songs of individual neighbors; they are sensitive to whether the songs they know come from the usual places or not. The birds thus appear to have a detailed knowledge of who belongs where in their vicinity and so are able to react with hostility to any change in the *status quo* of territorial arrangements.

The playback technique was used also to investigate the features that convey identity in the territorial singing of white-throated sparrows (Falls, 1963, 1969).

Falls compared the effects, in playback tests, of varying different features of the recorded songs. From these comparisons he made the observations that ". . . absolute pitch and possibly pitch change between notes are important for individual recognition" (Falls, 1969, p. 222). The conclusion is consistent with the fact that pitch characteristics are more constant than other features of the song of an individual but sufficiently different between individuals to provide a means of identification commensurate with the number of individuals that a male white-throated sparrow appears to be capable of recognizing.

The use of individual characteristics of voice for recognition in vocal communication between the members of a mated pair of birds has been investigated with playback of recordings in studies of antiphonal singing in the *Laniarius* shrikes (Grimes, 1966; Hooker and Hooker, 1969). When the duet of a neighbor was played back, both members of a pair approached the speaker and sang antiphonally with one another but apparently not with any precise temporal relationship to the playback song. When the component contributed by one of the birds of a pair to its duet was played, the other bird of the pair answered with its component, the "quoted" bird remaining silent (Hooker and Hooker, 1969). Similarly Watson (1969) has found in the eastern whipbird that a female will respond antiphonally to playback of the song of her mate, but to playback of songs of neighboring or distant males she shows either no response or "aggressive territorial defense." These results are consistent with the other evidence for believing that individual characteristics of duets enable the members of a pair to distinguish one another by ear from other birds in the vicinity.

2. Experiments on Young

Tschanz (1965, 1968) has used both successive and simultaneous playback of adult calls to investigate whether guillemot chicks can recognize the voices of their parents. The results are clear and positive. The chicks called, turned toward the speaker, approached it, snuggled up against it, and pecked at it (a feeding or food-begging response), when the sounds coming from it were "Lockrufe" of the parents, but when the calls were those of other adults the chick went into or remained in hiding. When calls of the parents were played from one speaker and calls of other adults from another the chick approached the speaker from which the calls of the parents came.

Similar results have been obtained in playback tests of chicks in black-billed gulls (Evans, 1970a), ring-billed gulls (Evans, 1970b) and Adelie penguins, *Pygoscelis adeliae* (D. H. Thompson, personal communication).

As well as providing a means of testing a bird's reactions to particular sounds, playback of recordings also offers a way of subjecting the bird to particular kinds of auditory experience. The technique has been used in studies of song development in several species of songbirds (e.g., Thorpe, 1958; Marler, 1967a; Konishi and Nottebohm, 1969). Tschanz (1965, 1968) has used it to investigate the development of recognition of the calls of the parents in guillemot chicks. He played a recording of "Lockrufe" of a guillemot to eggs in an incubator for 1–3 hours a day during the last 3 or 4 days prior to hatching. The chicks were later given choice tests in which the calls that had been played to them prior to hatching came from one speaker and similar calls from a different adult came from the other speaker. These chicks approached the calls they had previously been exposed to, whereas other incubator-hatched chicks, which had not experienced any calls prior to hatching, showed ambivalence, turning from one speaker to the other as though they were equally attracted to both. According to Tschanz, the guillemot chick's experience of hearing its parents' calls, prior to and just after hatching, changes a readiness to respond to any call into selective re-sponsiveness to particular calls: those of the parents. He suggests that this restriction of responsiveness is a consequence of the fact that the times when the unhatched chick is most active in the egg tend to coincide with calling by its own parents but not with the calling of other adults. The fact that, for a chick in the egg, the calls of its own parents will be louder, because they come from closer at hand, than the calls of other adults, may also be involved. [Evans (1970) found, in tests in which "mew calls" were played back to ring-billed gull chicks, that the louder the calls were played the stronger the approach they evoked.]

The functional value of such early development of parent recognition in the guillemot would appear to be that it prepares the young for the social congestion encountered from hatching onward. Guillemots breed on cliff ledges. The degree of crowding is such that, unless there were means to counteract it, the probability of members of a family becoming separated from one another would be very high. Given such a situation one must suppose either that care of offpsring is carried out indis-criminately by adults, without regard to family relationships, or that there must be some way by which the members of a family can recognize and find one another. There is no evidence of indiscriminate care of

young in guillemots. It appears that if the chicks had no way of recognizing their parents at hatching their chances of survival would be low. The
selection of the auditory modality for this function may be, in part at
least, a consequence of the fact that a chick inside an egg can hear its
parents but cannot perceive them visually or in other ways that would
allow the situation at hatching to be anticipated by development of
individual recognition of the parents prior to hatching.

To some extent at least, the conditions that make for individual
recognition by ear in the guillemot are present for other colonially
breeding species as well, but in less extreme forms in most cases. The
degree of social crowding to which guillemots subject themselves during
breeding is unusual in birds. We might therefore expect that, although
chicks may recognize the voices of their parents in colonial species in
general, the manifestations of this capacity differ in some respects from
what has been found in the guillemot. The recognition has been demonstrated in the colonial species that have been tested so far. The laughing
gull (*Larus atricilla*) provides an example of such a species in which
spacing out during breeding is much greater than it is in the guillemot.
This species has been the subject of my research on individual recognition of voice during the last few years.

III.   Individual Recognition of Voice in the Laughing Gull

The laughing gull colony in which my work has been carried out is
situated on low-lying marshy islands, surrounded by tidal channels, in
the Brigantine National Wildlife Refuge, a Federal reserve on the East
Coast of the United States. As in other species of gulls both members of
a pair share in the incubation of the eggs and care of the chicks. The
clutch size is usually three. General accounts of the breeding and social
behavior of the species will be found in Bent (1921) and Noble and
Wurm (1943).

Individual recognition of voice has been investigated in two contexts:
interactions between adults during the incubation period; interaction
between parents and young during the period between hatching and
fledging of the young. The study began with field observations. I found
that with certain of the calls of adults I could recognize individuals by
ear, and I got the impression that the adults and chicks could do likewise. The next steps were recording and spectrographic analysis of the
calls and the carrying out of playback experiments. The work of analysis
and experiment is far from complete so what I shall present here is, to
some extent, a progress report.

A. Individual Characteristics in the Calls of Adults

Of the several different kinds of call in the adult gull's repertoire only one, the "long-call," was to my ear so individually characteristic as to provide an easily learned and reliable means of identifying individuals. One other call, which I shall refer to as "ke-hah," was also individually distinct enough for me to be able to make consistently correct identifications with it in a few cases.

Long-calls, ke-hah calls, and other calls of a large number of gulls have been recorded on tape, and sonagrams were made of the recordings. Analysis of these is incomplete, but I have some preliminary results that I shall include here because they form a consistent story with some of the results of the playback experiments.

The long-call of the laughing gull consists of at least three distinct parts: a series of "short notes," a series of "long notes," and a series of "head-toss notes" (so named because a bird accompanies each note with throwing back its head). Sonagrams of long-calls are reproduced in simplified form in Fig. 1. They illustrate the variations that can be found in a sample of long-calls. For each part of the call there can be variation in the number of notes, in the durations of the notes and the durations of the intervals between them, in the "shapes" of the notes, and in their harmonic spectra. Long-calls are usually introduced by one or more disyllabic notes similar or identical to ke-hah notes; but sometimes long notes precede the short notes. Long notes and head-toss notes can also occur in isolation from the other components of the long-call.

Some of this variation is associated with differences in the context of occurrence. For example long-calls that occur as the overtures of "meeting ceremonies" (see Tinbergen, 1959), during pair formation and nest reliefs, typically have longer strings of long notes (between four and eight notes) than do long-calls uttered in other situations such as that of an incubating gull responding to gulls flying overhead (between one and three notes). Head-toss notes are usually missing from the long-calls of flying gulls, and their number in the long-calls of gulls on the ground appears to depend upon the proximity and status of other gulls in the vicinity, although the rules that govern their occurrence have yet to be made clear. If one regards an isolated string of long notes, with or without terminal head-toss notes, as a form of long-call, then whether or not a long-call includes short notes is almost certainly also decided by context or motivation, but again the deciding factors remain to be discovered.

For those long-calls in which short notes occur, however, there ap-

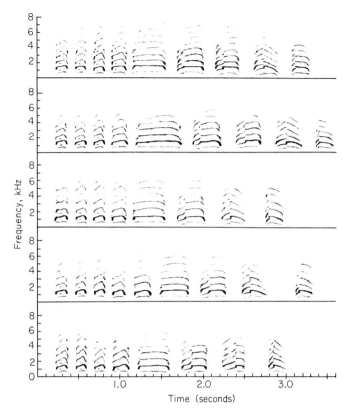

F<small>IG</small>. 1. Sonagrams of long-calls of laughing gulls. (a) Five calls by the same gull.

pears to be no variation associated with context in the short note part. Although the short note parts differ markedly between the long-calls of different individuals, they show little variation in the long-calls of a single individual, irrespective of the context of occurrence. The number of short notes, the "shapes" of short notes, and the durations of short notes and of the intervals between them are individually characteristic. The variation in these characteristics between individuals and their constancy within the calls of an individual are illustrated in Fig. 1. The pitch or timbre of the call as a whole may also be individually characteristic, but this impression is based more on the experience of identifying individuals by ear than on examination of sonagrams. Which of these individual characteristics in the long-call is of significance to the gulls remains an open question, the answer to which must await experiments along the lines of those carried out by Falls (1963, 1969) and Bremond

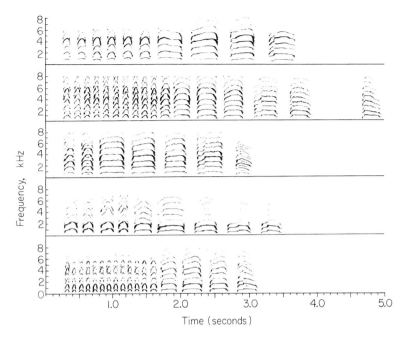

FIG. 1.(b) Calls of five other gulls.

(1967). In the meantime I submit that the long-calls of laughing gulls are individually characteristic and distinct enough for them to provide the gulls with a means of individual recognition.

Ke-hah calls are so named because they are abrupt and disyllabic to the ear. The variations that one can hear are in pitch and length of the two syllables relative to one another: the first syllable may be higher or lower or about the same pitch as the second; it may be longer, shorter or about the same length as the second. These characteristics are sufficiently different between individuals and sufficiently invariant in the calls of a single individual for one to identify individuals by ear. Comparisons of sonagrams (Fig. 2) bear out these impressions, and also show variation in the "shapes" of the syllables and in the duration and form of the transition from the first to the second, which is also individually characteristic.

Ke-hah calls typically occur in strings. The repetition rate and number of calls in a string vary, but this variation appears more likely to be a function of context and motivation than a product of individual idiosyncracy.

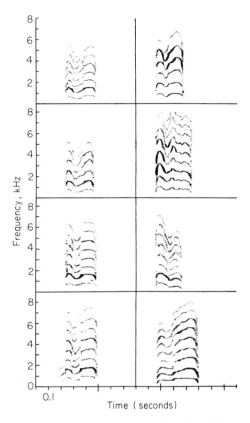

Fɪɢ. 2. Sonagrams of ke-hah calls of laughing gulls. The calls in the first column were by the same gull. The calls in the second column were by four other gulls.

Another call which is prominent in a variety of social contexts is a soft, smooth, drawn-out note with a plaintive quality that I refer to as "crooning" [cf. the "mew call" described for several other species of gulls (e.g., Tinbergen, 1953, 1959; Evans, 1970a,b)]. This call of the laughing gull varies considerably between utterances by the same bird, even between calls in a single string. In some cases the sonagram lines (Fig. 3) are straight and level, in others the line rises steadily until just before it vanishes, in others there is a step in the line which corresponds to a yodellike break in the sound; the durations of the calls vary, and so do the durations of the intervals between successive calls in strings. This variation in the crooning calls of a single individual overlaps with variation in the crooning calls of other individuals. As yet, individual charac-

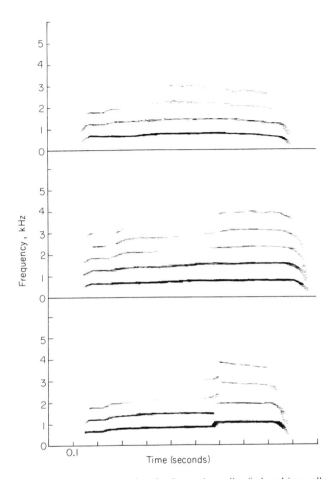

Fig. 3. Sonagrams (narrow band) of crooning calls of a laughing gull.

teristics in crooning calls have not been detected. Other types of calls in the laughing gull repertoire have so far not been examined.

B. Individual Recognition of Voice by Adults

Field observation provided numerous instances of an incubating gull reacting to the calls rather than the appearance of its approaching mate, e.g., when the mate was hidden from the view of the sitting bird by vegetation or the bank of a stream, or when the sitting bird was in the "tucked" position with its eyes closed.

Playback tests were conducted in the gullery in order to see whether incubating gulls would react differently to long-calls or ke-hah calls of different individuals. Pilot tests were carried out in 1967, the results of which led to more carefully designed tests made in 1968. I shall describe only the latter.

Two sets of test tape-recordings were prepared from recordings made in the gullery: a set of long-call tapes and a set of ke-hah tapes. A test tape consisted of four 1-minute sections on which calls were recorded, the sections being separated by 1-minute silent intervals. There was also 1-minute of silence at the beginning and end of the tape. Each record section contained calls of a different gull: two of the gulls represented were at neighboring nests in one part of the gullery, the other two were at neighboring nests in another part of the gullery, the two parts of the gullery being several hundred yards apart. The tape was made up so that a section from one part of the gullery alternated with a section from the other. In the long-call tapes each section consisted of four equally spaced long-calls of the same bird; in the ke-hah tapes each section consisted of between eight and ten variably spaced ke-hah calls of the same bird.

A test was arranged so that the gulls at two nests — the nests from which two of the sections on the test tape had originated — were tested at the same time. A portable loudspeaker was placed in a position that formed an isosceles triangle with the two nests. The speaker was aimed at the midpoint of the base of the triangle (the line between the two nests). The nests were between 3 and 5 yards apart and the distance from the speaker to the nests was about 7 yards. The speaker was connected by a 50-foot cable to a tape recorder which was operated from inside a hide. Volume controls on the tape recorder and speaker were set so that the loudness of the calls on the playback was, to my ears, in the range of naturally produced calls and the same as that which had been found adequate to produce response in the pilot tests.

Since, in each test, the test tape contained calls recorded at both nests and calls from nests in another part of the gullery, each bird tested heard calls that had originated from its own nest, calls that had originated from a neighbor's nest, and calls that had originated from what I shall refer to as foreign nests. Each test tape was used twice: once in each of the two places in the gullery at which the calls on it were recorded.

Unfortunately, it was not always possible to tell whether the bird on a particular nest during a test was the bird whose call had been recorded for the test or its mate. However, individual identification of the birds tested was made in sufficient of the tests for a number of them to be classed together as tests in which the bird on the nest was played the calls of its mate. The results of these tests are summarized in Table I.

TABLE I

RESPONSES OF INCUBATING LAUGHING GULLS TO PLAYBACK OF CALLS OF
THEIR MATES, NEIGHBORS, AND STRANGERS[a]

| | Playback | | | | | |
| | Long-calls | | | Ke-hah calls | | |
Responses	Mate	Neighbor	Stranger	Mate	Neighbor	Stranger
Resettling	6	0	0	2	0	0
Calling	9	0	0	8	0	0
Flying and collecting	3	0	0	0	0	0
Total responding	11	0	0	9	0	0
Total tested	15	15	15	12	12	12

[a] The scores are numbers of individuals.

The pilot tests had indicated that the responses to be expected in these playback tests were calling (either ke-hah or long-calls), rising and re-settling, or flying from the nest followed by return after collection of nest material. These are all responses that are commonly shown by the bird on the nest when its mate arrives for nest relief. Other responses, such as adoption of the thin-necked "upright" posture which is indicative of readiness to flee (see Tinbergen, 1959), occasionally occurred during playback in the later series of tests, but these, and anything other than quiet sitting during the silent intervals, were so rare that they have not been included in the table. As the table shows, the only clear responses in the tests were to playback of the calls of the mate. Long-calls and ke-hah calls did not differ in their effects, except that long-calls elicited flying from the nest and collecting in some cases but ke-hah calls did not.

Among the tests not included in Table I there were some in which it was known that the bird on the nest was played a recording of its own voice. In a few of these the bird responded to its own calls, in the ways the other birds responded to the calls of their mates. All were tests with long-calls; and in those for which sonagrams of the calls of male and female could be compared, the two birds were remarkably similar in the number, durations, and spacings of their short notes. However, the data are too few to permit the conclusion that a gull will respond to playback of its own long-call only if its own long-call resembles that of its mate. Nor can it be said, as yet, how common are pairs in which the long-calls of the two birds are similar.

Some observations of the pair-formation period indicate that a male will tolerate the presence of a female longer if the short notes of her long-call match his than if there is no match. Courtship display includes a performance in which the two birds utter long-calls side by side and in unison. I have recordings of such performances, in some of which the short notes of the two calls are in almost perfect synchrony. This synchrony is possible, obviously, only if durations of the short notes, the intervals between them, and their number in the calls of one bird are the same as those in the calls of the other. It may be that in unison calling the birds can, at least sometimes, adjust to one another's calls to achieve matching, or that unison calling is, so to say, a means by which the birds test whether their calls match. It is an open question whether birds with similar calls are paired because their calls are similar, or have similar calls because they are paired. Here too the data at present available are too fragmentary to permit one to do more than draw attention to interesting possibilities.

However, the playback tests leave no doubt that an incubating gull can distinguish the long-calls and ke-hah calls of its mate from those of other gulls. They might also suggest that gulls do not distinguish the calls of neighbors from those of strangers. But negative results are rarely conclusive, and, in this case, it would be premature to conclude that individual recognition of voice in adults occurs only between the members of a mated pair. For one thing the incubation situation is only one of a variety of contexts in which long-calls and ke-hah calls are used. For another, even in the context of incubation, individual recognition of voices other than that of the mate may be unexpressed in overt behavior unless the vocal signals are accompanied by visual stimuli which might indicate the course of action that a calling bird is pursuing. Incubating gulls are less likely to call in response to the ke-hah calls or long-calls of a flying gull if it is maintaining a steady course than if it is showing signs of landing nearby; and in the latter case, as a rule, less response is shown if the landing bird is a local nest owner than if it is not (Beer, personal observation). Another possibility is that different versions of the same kind of call may encode different "messages" (in the sense of Smith, e.g., 1968), only some of which provoke response in a sitting gull. All the calls in the test tapes used in the playback experiment were recorded from incubating birds that were, for the most part, responding to flying birds. I have already pointed out that certain features of the long-call vary with the context of occurrence. If, then, the version of a call uttered by an incubating gull to a gull overhead does not, by itself, provoke response in other sitting gulls, the question remains open whether the individual characteristics of the call are used by the gulls to distinguish neighbors from strangers. Further research is required, therefore, to examine the possibility of interaction between auditory and visual stimuli in individual recognition between adult gulls [e.g., using combinations of models and playback, as in the experiments of Stout *et al.* (1969)], and to elucidate the significance of variability, other than individual variation, in the calls [e.g., "message-meaning" analysis along the lines of Smith (1968)].

C. Recognition by Young of the Voices of Their Parents

Nice (1962) described laughing gull chicks as semiprecocial. They emerge from the egg with down feathers and open eyes but with limited powers of locomotion. During the first few days post-hatching they are dependent on their parents for warmth, and they are fed by their parents until after they are fledged, although they show some independent feeding beginning at about the second week.

Until the third or fourth day post-hatching the chicks rarely leave the nest, and one of the parents remains on the nest almost all the time, either brooding the chicks or standing over them. The parents feed the chicks by regurgitating food from the crop. Regurgitation appears to be stimulated by a chick's pecking at its parent's bill. Such pecking by the chick is elicited by the stimulus presented to the chick when the parent bends its head down and so brings its red bill vertically into the chick's visual field (Hailman, 1967). But a parent often croons during this bill presentation and also when it presents the regurgitated food at the tip of its bill. This crooning appears also to stimulate the pecking and feeding responses of the chick. At this early stage in the chick's life, crooning is the only call that the parents clearly direct at the chick and use to manipulate its behavior.

As a chick's powers of locomotion improve, its range of movement progressively extends beyond the immediate confines of the nest. On the third or fourth day post-hatching, parent gulls can sometimes be seen leading or luring their chicks on excursions away from the nest. A parent begins by holding its bill down and crooning as at the start of a typical feeding sequence, but as the chicks start responding the parent walks slowly from the nest, still holding its bill down and crooning. The chicks follow, calling and attempting to peck at the parent's bill, and thus they are led away from and then back to the nest, where they are eventually fed. To begin with such excursions may extend no more than a few inches from the nest rim. A day or so later they may extend several feet.

By the fifth or sixth day post-hatching the parents often land several yards away from the chicks and call to them from there. Now ke-hah calls are added to crooning or used instead of crooning until the chicks make bodily contact with the parent. The response of the chicks is usually vocalization and approach, followed by food-begging and feeding.

Up until about the eighth day post-hatching, chicks are rarely left unattended by at least one of their parents. But as the chicks get older they are left without the company of either parent for progressively longer periods, so that, by about the third week, many of the visits of a parent to its chicks last no longer than the few minutes it takes to feed the chicks. A parent returning to its chicks usually announces its arrival by long-calling as well as ke-hah calling, so that by the time the chicks are being left alone for periods, long-calling becomes as regular a prelude to feeding as ke-hah and crooning.

As the time that the parents spend with the chicks decreases, the range of the chicks' movements increases, bringing with it an increase in the frequency of encounter with adults other than the parents. Such encounters often result in a chick's being threatened, attacked, and forced to flee when outside its family territory; within the territory it is more likely that the chick attacks and drives off the adult. Such hostile interactions are rare between a parent and its chicks, irrespective of their place of occurrence. Thus, at least by the time chicks are being left alone by their parents, situations arise in which it is in a chick's interests to distinguish its parents from other adults, and in which such discrimination appears to be shown by the chick. The individual characteristics in the calls given by adults in these situations suggested a means by which the chicks could make the discriminations. Playback experiments have been carried out to investigate the possibility that laughing gull chicks can distinguish calls of the parents from the same kinds of calls in other adults (Beer, 1969, 1970a,b).

1. Recognition of the Voices of the Parents

In pilot tests it was found that a chick captured in the field at about 6 days of age post-hatching and tested indoors would vocalize and approach the source of sound in response to playback of a recording which contained crooning calls, ke-hah calls, and long-calls by one of its parents. In a later experiment, chicks between 6 and 12 days post-hatching were captured in the field and tested indoors with recordings similar to that found to be effective in the pilot tests.

Each chick was tested with two such recordings: one contained calls of one of its own parents; the other contained calls of a parent gull nesting in a distant part of the gullery [for details of experimental procedure, scoring and results of this experiment see Beer (1969)]. It was found that the chicks vocalized, oriented toward the sound, approached the sound, and showed continuous locomotion in response to the calls of their own parents but not to the calls of foreign parents. In response to the latter some of the chicks withdrew from the source of sound and crouched in silence. The experiment thus showed that laughing gull chicks in the age range tested can distinguish the voices of their parents from those of gulls from a different part of the gullery.

However, this experiment did not provide a conclusive demonstration that chicks recognize the voices of their parents. The differences in a chick's reactions to the two selections of calls played to it could have been based simply on familiarity, since in one case the calls were of a bird it had heard before and in the other case the calls were of a bird which, in all probability, it had never heard before. To carry the investigation further, the experiment was repeated with another group of chicks, captured in the field between 6 and 8 days post-hatching, but with the difference that each was tested with a recording of calls of one of its own parents and a similar recording of calls of a parent at a neighboring nest situated no farther than 15 feet away from its own. The chick was thus tested with recordings of two birds, the calls of both of which must have been familiar to it.

The results of this experiment were in close agreement with the experiment in which recordings of calls of foreign gulls were used: the chicks called, oriented toward the sound and approached it, and showed continuous locomotion when the recording of their own parents was played; they tended to orient away from the sound and withdraw from it, and some of the chicks crouched, when the recording of the neighbor was played (Beer, 1970a). Thus laughing gull chicks between 6 and 8 days post-hatching react to the calls of neighbors in the same way in which they react to the calls of strangers: both are distinguished from the calls of a chick's own parents, which alone, in the absence of the usual visual stimuli, can induce the chick to show the kind of behavior it reserves for its parents in the field. Two questions immediately arise: (a) at what age and as a consequence of what kinds of processes is recognition of the voices of the parents established? (b) on which of the different kinds of calls does the chick base its recognition?

2. Hand-Raised Chicks

An initial attack on part of the first of these questions was included in the experiment just described. At each nest from which the parent-raised subjects were taken, one of the eggs was removed and placed in an incubator during the last third of the incubation period. The chicks hatched from these artificially incubated eggs were hand-raised to between six and eight days post-hatching, when they were tested with the same recordings as their age-matched parent-raised siblings. In every case the hand-raised chicks fled from the sound and crouched in silence as far from it as they could get. They not only failed to show any recognition of the voices of their parents; they showed no filial response whatever to any of the calls. Experience prior to the age of 6 days post-hatching can therefore affect a laughing gull chick's responses to the calls of adults.

3. Age Differences

Effects of differences of experience or age were suspected in the earlier experiment (parent versus foreign calls) in which the ages of the chicks ranged from 6 to 12 days post-hatching. In the tests in which the recording of a chick's own parent was played, the measures of orientation toward the sound, approach to the sound and vocalization were negatively correlated with age of the chick (Beer, 1970b); i.e., the older a chick was the less positive response it showed to the same kind of selection of its parents' calls. This effect was investigated further in the later study. For comparison with the 6- to 8-day-old chicks, a group of younger chicks (1–3 days post-hatching) and a group of older chicks (12–28 days post-hatching) were selected from the field and tested, like the other chicks, with playback of parent and neighbor calls. The chicks of the early group (1–3 days) were thus tested at the stage when they were still more or less confined to the nest and the immediate proximity of parental company; the chicks of the middle group (6–8 days) were tested at the stage when their range of movement was extending beyond the immediate vicinity of the nest and they were being induced to approach their parents from a distance, but parental attendance on the chicks was still virtually continuous; the chicks of the late group (12–28 days) were at the stage when they were left unattended by their parents for periods, and their range of movement was bringing them into contact with adults other than their parents.

Some of the results are summarized in Fig. 4. Contrary to expectation, even the youngest chicks showed recognition of the voices of their parents: they oriented toward the sound and vocalized more often when the calls of their own parents were played than when the calls of neighbors were played. But, in contrast to the chicks in the middle and late groups, the chicks in the early group showed strong positive response

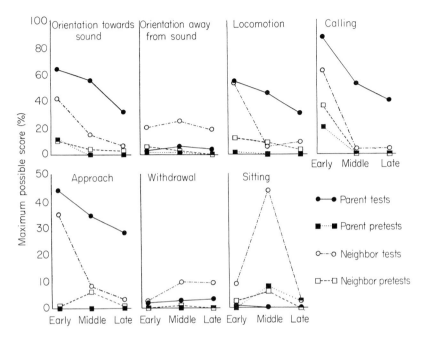

Fɪɢ. 4. Results of three age-groups of laughing gull chicks tested with playback of calls of their parents (parent tests) and calls of adults from a neighboring nest (neighbor tests). Early group: chicks 1–3 days post-hatching ($N = 8$); Middle group: chicks 6–8 days post-hatching ($N = 15$); Late group: chicks 12–28 days post-hatching ($N = 11$). The points are medians converted to percentages of the maximum possible score for each response measure. Pretests: 5-minute periods in which the chicks were observed without playback, preceding the 5-minute tests with playback.

to the calls of neighbors as well as to the calls of their own parents: in both sets of tests the scores for approach to the sound, orientation to the sound, locomotion, and vocalization were high compared to the other two groups. Thus individual differences of voice, although they can be detected by a 1- to 3-day-old chick, are not sufficient to deter the chick from responding positively to the calls of an adult that is not one of its own parents. By the time it is 6–8 days old, however, the chick shows virtually no positive response to calls of an adult not its parent; if anything the chick shows withdrawal and crouching in response to such calls. There were indications of a slight decline from the early to the middle group in positive response to playback of calls of a chick's own parent, but the only statistically significant difference between the scores for these two groups was in vocalization of the chicks. The chicks in the late group scored significantly lower than the chicks in the two younger groups on all measures of positive response in the parent tests: approach to the sound, orientation to the sound, locomotion, and vocalization.

To sum up the conclusions so far: although evidence of recognition of the voices of the parents can be found in chicks as young at 24 hours post-hatching, this discrimination sharpens markedly between the fourth and sixth day; and older chicks tend to respond less to playback calls in general than do younger chicks.

4. Differences in the Effects Produced by the Different Adult Calls

There were some indications in the earlier experiment that the negative correlations between the age of the chicks and their positive responses to calls of the parents might have been a reflection of change in the roles of the three different types of adult call in the control of a chick's filial responses as the chick grows older. The question of whether there are differences in the way a chick reacts to the three types of adult call was examined in an indirect way in the three age-groups of the later study.

Each tape recording used in the tests included instances of all three types of call: crooning, ke-hah, and long-call. One such recording was selected for each chick tested. It served as the chick's parental tape and as the neighbor tape for another age-matched chick, whose parental tape, in turn, served as the neighbor tape for the first. Effort was made to make the tape selections as similar as possible—e.g., as far as the numbers of each of the different kinds of calls were concerned—but some variation between tapes was unavoidable. However this variation provided ways of comparing the influences of the three different types of call on the behavior of the chicks: one could ask the question: to what extent and in what ways was variation in the behavior in the tests association with variation in the test recordings?

The results of the playback tests of the three age-groups were accordingly subjected to three kinds of analysis: correlation analysis, sorting of the tests according to the kind of playback call immediately preceding the initiation of positive response by the chick, analysis of temporal association between playback calls and chick behavior.

The correlation analysis was made on the numbers of each of the three types of call played in each test and the corresponding scores for each of the behavioral measures. Spearman rank correlation coefficients (Siegel, 1956) were thus worked out for each of the three sets of parent and neighbor tests (the early group, the middle group, and the late group). Some of the correlation coefficients were statistically significant ($p < .05$), and they suggest a changing role for the long-calls of the parents as a chick gets older. In the early group, the more often long-calling occured the more often a chick was oriented away from the speaker in both the parent tests ($r_s = .68$, $p < .05$) and neighbor tests ($r_s = .77$, $p < .05$). In the middle group, both orientation away from the speaker ($r_s = .52$, $p < .05$) and withdrawal ($r_s = .64$, $p < .01$) were positively correlated with long-calling in the neighbor tests, but long-calling showed no significant correlations in the parent tests. In the late group, withdrawal ($r_s = .68$, $p < .05$) and crouching ($r_s = .70$, $p < .05$) were positively correlated with long-calling in the neighbor tests, but in the parent tests orientation toward the speaker ($r_s = .85$, $p < .01$) and vocalization ($r_s = .87$, $p < .01$) were positively correlated with long-calling. It thus appears that long-calls tend to promote negative responses in a very young chick irrespective of whether the calls are given by one of the chick's own parents or another gull, this tendency persisting for the long-calls of adults other than the parents but not for the long-calls of the parents, which the chick eventually comes to react to with positive responses.

The significant correlations for ke-hah calling were few but consistent with the pattern seen in long-calling: ke-hah was positively correlated with a negative response (orientation away from the speaker) in the parent tests of the early group ($r_s = .70$, $p < .05$), and positively correlated with a positive response (locomotion) in the parent tests in the late group ($r_s = .81$, $p < .01$). The significant correlations for crooning did not conform to the pattern in long-calling and ke-hah: crooning was positively correlated with a positive response (approach) in the neighbor tests of the early group ($r_s = .76$, $p < .05$), and with a negative response (crouching) in the neighbor tests of the late group ($r_s = .85$, $p < .01$). The only significant correlations of crooning in parent tests were in the late group: negative correlation with withdrawal ($r_s = -.68$, $p < .05$) and positive correlation with vocalization ($r_s = .66$, $p < .05$).

To sum up: a laughing gull chick's response tendencies to long-calls and ke-hah calls appear to shift from negative to positive for the calls of its own parents, but remain negative for the calls of gulls other than the parents, as the chick increases in age. Response tendencies to crooning, on the other hand, appear to shift from positive to negative for the calls of adults other than the parents, but not for those of the parents.

To get an indication of the type of call initiating positive response in the tests I took each of the three types of playback call and, for each, divided the tests in which positive responses occurred into those in which the initiation of positive response was preceded within 15 seconds by occurrence of the calls and those in which this was not so. My index of the initiation of positive response was the first occurrence of vocalization by the chick in a test. I made this choice because vocalization is almost always the first response made by a chick in response to arrival of one of its parents in the field; in the tests it almost always preceded the first clear orientation to the speaker and approach by a few seconds; and it was the most distinct item of behavior recorded in the tests, at least as far as time of occurrence was concerned. Also all of the behavior items I have lumped together as positive responses to playback—approach, orientation toward the sound, locomotion, and vocalization—occur together in the field and were highly correlated with one another in the tests.

The distributions of the tests which resulted from this analysis (Table II) revealed a pattern consistent with that found in the correlation anal-

TABLE II

PLAYBACK TESTS OF LAUGHING GULL CHICKS[a]

Test type	Group	Tests in which first chick call preceded by—			Total tests in which the chick called
		Crooning	Ke-hah	Long-call	
Parent	Early	16	10	4	16
	Middle	15	17	13	28
	Late	8	13	11	17
Neighbor	Early	13	6	7	16
	Middle	6	11	6	15
	Late	1	5	0	5

[a] The tests in which the chick called divided according to the kinds of playback calls that preceded (within 15 seconds) the first call of the chick.

ysis. Comparison of parent tests of the early and late groups showed that the proportion of tests in which the first call of the chick was immediately preceded by long-calling was significantly higher in the late group ($\chi^2 = 5.32$, $p < .05$). In the neighbor tests, on the other hand, the proportion was, if anything, lower in the late group, but the difference was not significant. The proportion was not significantly different between parent and neighbor tests of the early group, but was significantly higher in the parent tests than in the neighbor tests in the late group ($p = 0.025$, Fisher exact probability test). The proportions for ke-hah calling did not differ significantly between early and late parental tests, but in the neighbor tests significantly more of the chicks uttered their first calls immediately after ke-hah in the late group than in the early group ($p =$ 0.05, Fisher test). However parent and neighbor tests did not differ significantly, as far as the proportions for ke-hah calling were concerned, in either the early or late groups. Crooning showed a pattern opposite to that of long-calling. The proportion of tests in which the first call of the chick was immediately preceded by crooning was lower in the late group than in the early group in both the parent tests ($\chi^2 = 9.31, p < .01$) and the neighbor tests ($p = .05$ by Fisher test). The proportions for parental and neighbor tests were not significantly different in either the early or the late groups. In the early group the first call by the chick was immediately preceded by crooning in every case in the parent tests (16/16) and in almost every case in the neighbor tests (13/16). In the late group, in those tests in which calling by the chick occurred, the first call of the chick was immediately preceded by either ke-hah or long-call in every case in the parent tests (17/17), and by ke-hah in every case in the neighbor tests (5/5).

It thus appears that positive response tends to be initiated by crooning when chicks are very young, but that by the time the chicks are about 12 days of age this function is taken over by ke-hah calling and long-calling of the parent and can sometimes be performed by ke-hah calling, but not long-calling, of adults other than the parents.

The analysis of temporal association between occurrence of playback calls and behavior in the tests involved sorting out the number of 15-second intervals in each test in which a particular type of call and a particular item of behavior occurred together, and then working out whether the resulting numbers were greater or less than chance expectation (for details of the method see Beer, 1970b). The analysis showed that for some combinations of call-type and behavior item, the numbers were significantly high ("positive association") and for others they were significantly low ("negative association"). Again a pattern emerged which was consistent with that found in the other two analyses. In the early and middle groups long-calling was less often accompanied by positive items (orientation toward the sound, locomotion, vocalization) than one would expect by chance, in both the parent and neighbor tests. In the late group, however, long-calling and positive items (locomotion, vocalization) occurred together significantly more often than by chance in the parental tests, but there was no evidence of such a shift in the neighbor tests. An earlier shift from negative to positive association with positive items was indicated in the results for ke-hah. In parent tests ke-hah calling was negatively associated with locomotion in the early group, positively associated with orientation to the sound and with vocalization in the middle group, and with vocalization in the late group. In the neighbor tests, however, negative association between ke-hah and locomotion persisted from the early to the middle group. Crooning showed positive association with positive items (locomotion, vocalization) in both parental and neighbor tests in the early and middle groups, but in the late group the results were mixed: negative association with locomotion and positive association with orientation to the sound in the parental tests; positive association with approach to the sound in the neighbor tests.

Each of the three types of analyses of the playback tests is less precise and less direct an attack on the question of whether a chick reacts differently to the different types of adult call than one might wish; but the conclusions they point to are in general agreement with one another, so that what they lack in exactness individually is to some extent compensated by their collective weight. The results of the three analyses converge in a pattern which suggests the following interpretation. During the first few days of a chick's life out of the egg, when it is more or less confined to the nest and has the constant companionship of one or other of its parents, its filial behavior is stimulated by crooning but tends to be suppressed by ke-hah calling and long-calling. When the chick is older, and its range of social contacts has enlarged to include adults other than its parents, the ke-hah calling and long-calling of its parents cease to deter the chick and acquire the power to elicit vocalization, approach, and so on by the chick, thus usurping or replacing part of the function earlier served by crooning. Ke-hah calling may make this transition earlier than long-calling (evidence from analysis of temporal association). Long-calls and ke-hah calls by adults other than the parents, however, continue to have negative effects on the behavior of the chick as it grows older, although the ke-hah calls, by themselves, may sometimes initiate positive response in older chicks. Crooning still retains some of its positive effect on filial behavior, but this is mixed with negative effects which suggest that, for older chicks, crooning requires the support of ke-hah or long-calling by one of the parents to contribute its part to parent-chick interactions.

This last point was tested by subjecting some of the older chicks to playback of two additional recordings of their own parents. One consisted of an almost uninterrupted string of crooning calls; the other consisted of an almost uninterrupted string of ke-hah calls (no such recordings of long-calls were available). The scores for orientation to the sound, approach, locomotion, and vocalization were high in the ke-hah tests but low in the crooning tests (Beer, 1970b). The results of the analyses of the other playback tests were thus confirmed. Unfortunately it was not possible to carry out such tests with younger chicks.

5. Hand-Raised Chicks Retested

That experience of the early or crooning stage is important for the development of a chick's responses to the calls of its parents in the later or ke-hah-long-call stage, was suggested by the results of a final series of tests. After they had been tested in the experiment described earlier, the incubator-hatched hand-raised chicks were placed back in the nests they had originally come from, together with their parent-raised siblings. Six of the thirteen hand-raised chicks were accepted by their parents and so survived (all of the parent-raised chicks were accepted back by their parents). After they had been with their parents for 6 days — i.e., for an amount of time equivalent to the time following hatching within which a parent-raised chick acquires individual recognition of its parents' voices — these chicks were retested with the same recordings as previously. Their parent-raised siblings were retested also. The hand-raised chicks no longer fled and crouched in response to the playback calls, but they still failed to show either recognition of their parents' voices or any positive response to the playback calls. They all virtually ignored the calls. Their parent-raised siblings, on the other hand, showed parent recognition and positive response within the range shown by the chicks in the late group of the experiment with the three age-groups.

My impression from observation of the hand-raised chicks when they were with their parents, was that they were cueing their behavior to the behavior of their siblings rather than the behavior of the parents. By keeping close to a sibling with normal experience, a hand-raised chick was on the spot when food was regurgitated by the parents (Beer, unpublished observations). It thus appears that if a chick is deprived of normal interaction with its parents during the first 6–8 days post hatching, recognition of the parents' voices and appropriate responsiveness to parental calls either fail to develop or take longer to develop, in a family situation, than they do in the normal course of events.

IV. Discussion

Thorpe (1968) has recently argued that efficiency and success in breeding in colonial birds in effect demand that the individuals of a family be able to recognize one another. The necessity might be questioned, for there are conceivable alternatives to individual recognition as a means to preserving social order and well being, such as those suggested by insect societies. But it is a fact that evidence of individual recognition between parents and young has been found in every colonial species in which it has been sought, with the single exception, to my knowledge, of the kittiwake *Rissa tridactyla* (Cullen, 1957). And even the kittiwake may be regarded as an exception that proves the rule, for in that species the young remain confined to the nest until they fledge (Cullen, 1957), so that, up to that time, homing to the nest is sufficient to enable a parent to find its young, and the young do not normally encounter adults other than their parents.

But the case of the kittiwake also makes the point that in comparing different species one is likely to find that features of social behavior may be intricately related to one another and to other aspects of the breeding biology. For example kittiwakes and guillemots have evolved quite different ways of utilizing cliffs as breeding habitats: the kittiwake has become specialized in the use of narrow ledges; the guillemot has become specialized in the use of wider ledges. The kittiwake solution obviates the need for individual recognition between parents and their young; the guillemot solution apparently necessitates such recognition from the time of hatching onward.

Colonial species nesting on flat ground face selection pressures different from those faced by cliff-nesting species. Some of the pressures favor spacing out, others favor aggregation; consequently the outcome can be viewed as a compromise or balance of opposing forces (see Kruuk, 1964; Patterson, 1965; Tinbergen *et al.*, 1967, on spacing out in black-headed gulls). The balance is apparently different in different species. For example, black-billed gulls nest closer together than do black-headed gulls, and there are a number of respects in which this difference appears to be reflected in differences in the breeding behavior in the two species, related to differences in their breeding ecology (Beer, 1966).

Black-billed gulls probably nest closer together than any other species of gull. Even so they space themselves farther apart than do guillemots, and they build nests within which the young remain for the first few days after hatching. It would thus appear that there is less urgency for a black-billed gull chick to be able to recognize its parents than for a guillemot chick. Evans (1970a) has conducted playback tests on black-billed gull chicks and found that mew calls of the parents were not discriminated from those of other adults before chicks were 3 or 4 days of age, which is about the age at which they abandon the nest and begin intermingling with the members of other families.

Laughing gulls nest considerably farther apart than do black-billed gulls, and their chicks are usually not faced with situations demanding recognition of the parents until they are at least a week old. Nevertheless laughing gull chicks as young as 24 hours post-hatching showed discrimination between recordings of calls of their parents and recordings of calls of neighbors in the playback tests I conducted. In spite of this discrimination, however, the chicks younger than 4 days post-hatching responded with vocalization, approach, and so on to playback of a neighbor's calls as well as to playback of the parent's calls. It was only with chicks 6 days and older that these responses to playback were restricted more or less exclusively to the calls of the parents.

As we have seen, this change in responsiveness in the playback tests of laughing gull chicks was associated with a change in which of the three kinds of calls stimulated the responses: in the younger chicks the responses were stimulated by crooning but ke-hah and long-calls had negative effects; in the older chicks ke-hah and long-calls of the parents rather than crooning stimulated the responses, but ke-hah and long-calls of the neighbors continued to have negative effects. There is at present no evidence that crooning calls are individually distinct enough to convey individual identity, but ke-hah calls and long-calls can be used to identify individuals by ear. Moreover, ke-hah calls and long-calls are probably superior to crooning calls in the extent to which they can be localized and thus enable a chick to orient its movements to birds it cannot see as, for instance, when it is in the thick grass which grows up around the nest sites. According to Marler (e.g., 1967b), there are three ways in which a sound can affect the two ears differently and thus convey the direction of its origin: high frequencies result in intensity differences; low frequencies result in phase differences; transient frequencies and abrupt stops and starts provide detectable differences in time of arrival of the sound. Ke-hah calls and long-calls possess all three of these direction giving features; crooning has only its relatively low pitch.

Thus by the time that ability to identify and localize the parents from a distance is demanded by situations in which laughing gull chicks find themselves, the filial responsiveness of the chicks to voice has been transferred from a call that is not a good indicator of either identity or locality to calls that are. As yet it is not possible to say whether the discrimination shown by the very young chicks was based on crooning or on ke-hah and long-calls. On balance I think the evidence is in favor of the hypothesis that, in the tests of the early group, the approach, vocalizations, and so on shown by the chicks were indiscriminant responses to crooning, the differences between the parent and neighbor tests being due to the ke-hah and long-calls of the neighbor having greater negative effects than the ke-hah and long-calls of the parent.

If the hypothesis is correct, then we have an interesting difference between laughing gulls, on the one hand, and black-billed gulls and ring-billed gulls on the other, for Evans (1970a,b) has found in both the latter species that chicks can recognize their parents by their mew calls. The mew call is almost certainly homologous and functionally analogous to crooning in the laughing gull. But in contrast to crooning in the laughing gull, mew calls in black-billed gulls and ring-billed gulls have individually identifying characteristics which are easily seen in sonagrams. A reason for this individual distinctness in mew calls, at least in black-billed gulls, may be that the nests are so close together that once the chicks are mobile an indiscriminant response to mew calls would be liable to lead a chick into a neighbor's nest. The distances between laughing gull nests, on the other hand, are great enough for a chick on or near the nest to experience the crooning of one of its own parents as louder than crooning of a neighbor; and this difference could be sufficient to keep a chick from attempting to respond to crooning by a neighbor, at least until the chick's mobility increases to the point when its wanderings are likely to take it close to a neighbor's nest, by which time ke-hah and long-calling have superseded crooning in the control of filial responses. Evans (1970b) found, in playback tests of ring-billed gull chicks, that the rate of approach increased with increase in the loudness with which mew calls were played. Tschanz (1968) also found with guillemot chicks that "by increasing volume, an uninteresting call is made more effective."

Another point of contrast between black-billed gulls and laughing gulls is that the development of mobility in black-billed gull chicks and the age at which they abandon the nest (Beer, 1966) are in advance of what one finds in laughing gull chicks. Black-billed gull chicks, together with their parents, depart the nest at about 4 days of age post-hatching to lead what could be described as a nomad existence in which they encounter the members of other families. At this age laughing gull chicks are still at the stage when their parents are using crooning calls to lure them on short excursions from the nest, i.e., before ke-hah or long-calling have taken over the role of crooning and before the nest site ceases to be the focus of family life. Suppose that there is a parallel shift in the responsiveness of chicks to the calls of their parents, in black-billed gulls and laughing gulls, and that the timing of this shift is about the same in the two species. Then the relatively early abandonment of the nest for life in a flock which occurs in black-billed gulls offers a plausible reason why the chicks should be able to recognize their parents by their mew calls in this species but not in the laughing gull. But this suggestion is obviously very tenuous and is offered in ignorance of how black-billed gull chicks react to the long-calls and other adult calls apart from crooning.

Another tenuous possibility is suggested by the role of the long-call in parent–chick interactions in the laughing gull. We have evidence that the chicks learn to recognize the long-calls of their parents, but there is no evidence that they react differently to the male and female parent, or that there is any functional reason why they should, since they are treated alike by both. If both members of a pair have the same identifying characteristics in their long-calls, then their chicks have to learn only one set of characteristics instead of two. If there were a selective advantage in thus having the learning problem posed for the chicks as simple as possible, one would predict that pairs in which identifying characteristics of the long-calls of the two birds were matched would be favored. Such matching has been found in some pairs (page 49).

The discussion so far has sought to illustrate how study of aspects of social interactions in a single species, and comparison with other taxonomically related or in some respects similar species, can reveal or suggest functional correspondences between selective responsiveness based on individual recognition of voice, the vocal signals involved, and the demands of different social and ecological situations. The concentration on vocal communication of identity, to the neglect of visual or other means of individual identification, in the studies that have been drawn on, was due mainly to the kinds of practical reasons referred to in the introduction. But vocal communication has obvious advantages in situations in which visual communication — as far as we know the only alternative form of communication at a distance available to birds — is liable to be hampered [see Marler's comparisons of the different sensory modalities as channels for social communication (e.g., Marler and Hamilton, 1966)], and this suggests a feature that might be common to the instances of the use of voice as a means of conveying individual identity.

The good evidence of individual recognition of voice in birds comes either from colonial species, such as those just discussed, or species that live in thick vegetation. Foliage obstructs vision but impedes sound to only a minor degree, so the advantage of vocal communication in forest or jungle needs no elaboration. In the colony situation its advantage probably lies in the fact that a sound is transmitted in all directions and can be received from all directions, whereas a visual signal usually has a relatively narrow field of projection and has to be looked at to be seen. A crowded community presents a situation in which visual scanning or search might require some time for location of a particular individual, during which a bird could, for instance, be exposing itself to the hostility of others. The omnidirectional characteristics of sound transmission and reception in birds probably allows for more ready detection of individuals in such circumstances, given vocalizations that are individually distinct enough and have the properties that facilitate localization.

This aspect of the function of vocal recognition between adult laughing gulls was illustrated by what can happen when it is not available. After a general alarm that puts most or all of the birds in a part of the gullery into flight, a gull has to find its way back to its nest without the aid of the voice of its mate to guide it. Landing and settling on the wrong nest are not uncommon in such circumstances, with the consequence that colony alarm flights are sometimes followed by a flurry of hostile interactions as the birds struggle to find and reclaim their nests.[2] The haste to settle on their nests, which no doubt reflects the fact that an exposed nest is liable to predation, apparently leaves some of the birds too little time to make a correct visual identification of their nests. But the advantage of vocal identification is also evident in numerous instances of gulls establishing contact when they are out of sight of one another, e.g., when a chick is in tall grass and is alerted to the arrival of one of its parents.

If individual recognition is important in the social life of nocturnal birds, or birds such as the *Collocalia* species which breed in crowded communities in the dark of caves, it must surely depend upon voice here also. Indeed, elaboration of voice for individual recognition and other communication functions may well be extreme in such species in which vision is so restricted as an alternative or support to the auditory channel. One looks forward to studies of such species, and also, for comparison, to studies of species in which vision might have been exploited for the purpose of individual recognition.

Of course individual recognition may be more important, more highly developed or more presistent in the social behavior of some species than in others, and in some it may be lacking altogether. In species in which social ties or relationships are of brief duration, and the individuals are otherwise solitary, individual recognition would be surprising. Even in species in which the pair bond is prolonged and there is extensive parental care of young, fidelity to place rather than fidelity to individuals could be sufficient to keep a family together. Some forms of territoriality would make this possible.

[2] Experiments in which nests were transplanted showed that laughing gulls will tolerate a shift of several feet; as long as there was an unoccupied nest in the vicinity of their nest site they would settle in it and carry on incubating the eggs. Transplantations of nests as a result of tidal flooding are not uncommon in the Brigantine gullery, so the looseness of the tie to the original nest site may be an adaptation to this feature of the nesting habitat (Beer, personal observations).

Thorpe, in discussing vocal mimicry and antiphonal singing in birds, has hinted that the use of vocalization for personal recognition, and hence the maintenance of pair and family bonds, may be more highly developed in species in which the use of vocalization as a means of territorial advertisement and defense is relatively slight (Thorpe and North, 1966; Thorpe, 1967). This suggestion encourages further comparative study of individual recognition of voice in gulls. Two of the species in which vocal recognition has been demonstrated experimentally are not territorial in the classical sense, or only weakly so: the black-billed gull (Beer, 1966; Evans, 1970a) and the laughing gull (Beer, personal observations). It would be of interest to know how individual recognition in these species compares with individual recognition in species such as the black-headed gull and herring gull which are more classically territorial (Kirkman, 1937; Moynihan, 1955; Tinbergen, 1953).

However it is already evident, at least in certain types of calls of the laughing gull, that functions as different as the conveying of hostility and the conveying of identity do not necessarily exclude one another but can be served by the same call. The long-call, for example, combines a variety of signal functions: e.g., threat in agonistic contexts, attraction of the opposite sex in courtship contexts and at the same time the signaling of individual identity and probably also species identity. As has been shown in other species [e.g., the brown towhee (Marler and Isaac, 1960b)], the different communication functions are probably served by variation in different parameters of the call, in combination with differences in context of occurrence.

I now turn to developmental aspects of individual recognition of voice. One has even less information on this matter than on the functional aspects, and most of it comes from the gull studies.

In addition to the coincidences that suggest a functional pattern in the changes in a laughing gull chick's responsiveness to the calls of adults, there is a further coincidence which suggests how the transition from response to crooning to response to ke-hah and long-call is brought about developmentally. For the very young chick, crooning is usually contiguous with its being fed. As the chick grows older, and has to approach its parents from a distance, ke-hah and then long-calls become regular preludes to the feeding sequences. Since this progression is coincident with the change in the chick's responsiveness to its parents' calls, it might be suggested that this change is a consequence of conditioned chaining of responses, with feeding as the terminal reinforcement (cf. Holland and Skinner, 1961).

Two other facts are consistent with this suggestion. First, in playback tests of hand-raised chicks the only positive responses ever shown were to the sound of human voice, i.e., the only sound played to them that resembled sounds that had regularly occurred at times when they were fed. Second, when these chicks were placed with their parents in the field, those that were accepted were still not responding positively to their parents' calls after 7 days of experience with them. These chicks had been deprived of experience of the early crooning stage. That they failed to develop appropriate responses to the calls of their parents may have been because they were without this early experience to build on. It has also been found in a study of the pecking response of herring gull chicks that pecking rate and "pecking preference" can be affected by conditioning to food reward (Hailman, 1967).

However Evans (1970b) found in his study of the ring-billed gull that chicks do not require food reinforcement to learn to discriminate between chicks they have been raised with and strange chicks or to develop approach responses to their parents. Also, of course, the initial tendency of gull chicls to peck at the parent's bill does not depend on prior experience of its consequences. And there is evidence that the effect of crooning on the laughing gull chick's behavior may initially depend upon experience prior to the first feeding. Impekoven (1969) played crooning calls to pipping eggs in an incubator, and when the chicks hatched she tested their pecking response to a red rod with and without accompaniment of crooning. The pecking rate was higher when presentation of the rod was accompanied by crooning than when it was not. Control chicks that were not exposed to crooning prior to hatching did not show a higher pecking rate when crooning was added to presentation of the rod. There is thus evidence that experience prior to hatching can influence the initial responsiveness of a laughing gull chick to the calls of its parents, but there is as yet no evidence that the chicks acquire recognition of the voices of their parents prior to hatching. Even though there appears to be much less urgency for a laughing gull chick to be able to recognize its parents than for a guillemot chick, the possibility that it can acquire such recognition while still in the egg cannot be ruled out, particularly in view of the fact that laughing gull chicks have been found to show recognition of the voices of their parents as early as 24 hours post-hatching, which is about a week before they can put the capacity to any use.

However Evans (1970a,b) has found that, in the black-billed gull and the ring-billed gulls, chicks do not recognize the mew calls of their parents until they are 4 or 5 days old. In his studies of the ring-billed gull he found evidence of progressive refinement in the chicks' selective responsiveness to social companions. He distinguished three stages: (1) an "incipient mobility" stage (1st–3rd day post-hatching) when the chicks are more or less confined to the nest but respond with approach to a wide range of stimuli; (2) a "restricted mobility" stage (4th–5th day post-hatching) during which they take relatively short temporary excursions away from the nest and manifest "species recognition" in their approach responses; (3) an "extended mobility" stage (5th day-fledging) in which they make permanent and relatively lengthy emigrations away from the nest and display individual recognition. By manipulating the early experience of hand-raised chicks Evans was able to find evidence of a "critical period" for the development of approach responses to conspecifics, and hence, presumably, for the development of individual recognition. As we have seen, the early development of social behavior in a laughing gull chick also consists of a progression each stage of which may depend, in part at least, upon experience had in the preceding stages, perhaps all the way back to prehatching. Deprivation of normal early experience, interruption of the continuity of the normal progression of social interactions, may thus lead to retardation or deviation in a chick's social development. Whether failure to acquire recognition of the parents in the normal way as a chick has consequences for the social behavior of the fully mature adult laughing gull remains to be seen. Investigation of this question might well add to the growing body of facts, from a variety of animals, indicating that ability to form normal social relationships in adulthood depends on formation of certain kinds of social relationships in early life.

Finally it has to be confessed that we have been no more successful at answering some of the questions about vocal communication in birds that Craig raised in 1918 than he was. But at least there has been a revival of interest in such questions, and the outcome may prove to be an Advance in the Study of Behavior.

Acknowledgments

The laughing gull work reported here was supported by Grants Nos. GM 12774 and MH 16727 from the U.S. Public Health Service and a grant from the Research Council of Rutgers University. It could not have been carried out without permission from the U.S. Fish and Wildlife Service to work in the Brigantine National Wildlife Refuge and to use a Federally owned building there as a Field Station. I should also like to express gratitude to the Refuge Manager and his staff for their hospitality and cooperation, to Miss Joan Colsey who took responsibility for the care of the hand-raised chicks, and to my colleagues on the laughing gull project, particularly Dr. Monika Impekoven, whose cooperation and council were of considerable help to me. I thank Dr. S. J. White for her helpful comments on the manuscript.

References

Abs, M. 1963. Field tests on the essential components of the European nightjar's song. *Proc. 13th Int. Ornithol. Congr. Cornell, 1962* pp. 202–205.

Anderson, A. H., and Anderson, A. 1957. Life history of the cactus wren. *Condor* **59**, 274–296.

Andrieu, A. J. 1963. Techniques used for the physical study of acoustic signals of animal origin. *In* "Acoustic Behaviour of Animals" (R.-G. Busnel, ed.), pp. 25–47. Elsevier, Amsterdam.

Armstrong, E. A. 1963. "A Study of Bird Song." Oxford Univ. Press, London and New York.

Beebe, W. 1918–1922. "A Monograph of the Pheasants." Witherby, London.

Beer, C. G. 1966. Adaptations to nesting habitat in the reproductive behaviour of the black-billed gull *Larus bulleri. Ibis* **108**, 394–410.

Beer, C. G. 1969. Laughing gull chicks: recognition of their parents' voices. *Science* **166**, 1030–1032.

Beer, C. G. 1970a. On the responses of laughing gull chicks to the calls of adults. I. Recognition of the voices of the parents. *Anim. Behav.* **18**, 652–660.

Beer, C. G. 1970b. On the responses of laughing gull chicks to the calls of adults. II. Age changes and responses to different types of call. *Anim. Behav.* **18**, 661–677.

Bent, A. C. 1921. Life histories of North American gulls and terns. *U.S. Nat. Mus. Bull.* **113**.

Bergman, G. 1946. Der Steinwälzer, *Arenaria i. interpres* (L) in seiner Beziehung zur Umwelt. *Acta Zool. Fenn.* **47**, 1–152.

Borror, D. J. 1960. The analysis of animal sounds. *In* "Animal Sounds and Communication" (W. E. Lanyon and W. N. Tavolga, eds.), Publ. No. 7, pp. 137–320. Amer. Inst. Biol. Sci., Washington, D.C.

Borror, D. J. 1961. Intraspecific variation in passerine bird songs. *Wilson Bull.* **73**, 57–78.

Borror, D. J., and Gunn, W. W. H. 1965. Variation in white-throated sparrow songs. *Auk* **82**, 26–47.

Borror, D. J., and Reese, C. R. 1953. The analysis of bird songs by means of a vibralyzer. *Wilson Bull.* **65**, 271–303.

Bremond, J.-C. 1963. Acoustic behaviour of birds. *In* "Acoustic Behaviour of Animals" (R.-G. Busnel, ed.), pp. 709–750. Elsevier, Amsterdam.

Bremond, J.-C. 1967. Reconnaissance de schémas réactogènes liés à l'information continue dans le chant territorial du rouge-gorge. *Proc. 14th Int. Ornithol. Congr., Oxford, 1966* pp. 217–229.

Brückner, G. H. 1933. Untersuchungen zur Tiersoziologie, insbesondere der Auflösung der Familie. *Z. psychol.* **128**, 1–120.

Busnel, R.-G. 1963. On certain aspects of animal acoustic signals. *In* "Acoustic Behaviour of Animals" (R.-G. Busnel, ed.). Elsevier, Amsterdam.

Buxton, A. 1946. "Fisherman Naturalist." Collins, London.

Cherry, C. 1966. "On Human Communication." M.I.T. Press, Cambridge, Massachusetts.

Collias, N. E. 1952. The development of social behavior in birds. *Auk* **69**, 127–159.

Craig, W. 1908. The voices of pigeons regarded as a means of social control. *Amer. J. Sociol.* **14**, 66–100.

Craig, W. 1918. Appetites and aversions as constituents of instincts. *Biol. Bull.* **34**, 91–107.

Cullen, E. 1957. Adaptations in the Kittiwake to cliff-nesting. *Ibis* **99**, 275–302.

Curio, E. 1959. "Verhaltensstudien am Trauerschnäpper." *Z. Tierpsychol.* **3**, 1–118.

Davies, S. J., and Carrick, R. 1962. On the ability of crested terns, *Sterna bergii*, to recognize their own chicks. *Aust. J. Zool.* **10**, 171–177.

Dilger, W. C. 1956. Hostile behavior and reproductive isolating mechanisms in the avian genera *Catharus* and *Hylocichla. Auk* **73**, 313–353.

Dircksen, R. 1932. Die Biologie des Austernfischers, der Brandseeschwalbe und der Küstenseeschawlbe nach Beobachtungen und Untersuchunger auf Norderoog. *J. Ornithol.* **80**, 427–521.

Dumortier, B. 1963. The physical characteristics of sound emissions in arthropoda. *In* "Acoustic Behaviour of Animals" (R.-G. Busnel, ed.), pp. 583–654. Elsevier, Amsterdam.

Evans, R. M. 1970a. Parental recognition and the "mew call" in black-billed gulls *(Larus bulleri), Auk* **87**, 503–513.

Evans, R. M. 1970b. Imprinting and the control of mobility in young ring-billed gulls *(Larus delawarensis). Anim. Behav. Monogr.* (in press).

Falls, J. B. 1963. Properties of bird song eliciting responses from territorial males. *Proc. 14th Int. Ornithol. Congr. Oxford, 1966* pp. 259–271.

Falls, J. B. 1969. Functions of territorial song in the white-crowned sparrow. *In* "Bird Vocalizations" (R. A. Hinde, ed.), pp. 207–232. Cambridge Univ. Press, London and New York.

Goethe, F. 1937. Beobachtungen und Untersuchungen zur Biologie der Silbermöwe auf der Vogelinsel Memmerstrand. *J. Ornithol.* **85**, 1–119.

Goethe, F. 1954. Experimentelle Brutbeendigung und andere bruthiologische Beobachtungen bie Silbermöwen *(Larus a. argentatus* Pontopp.). *J. Ornithol.* **94**, 160–174.

Gottlieb, G. 1965. Imprinting in relation to parental and species identification by avian neonates. *J. Comp. Physiol. Psychol.* **59**, 345–356.

Greenwalt, C. H. 1968. "Bird Song: Acoustics and Physiology." Smithsonian Inst. Press, Washington, D.C.

Grimes, L. 1965. Antiphonal singing in *Laniarius barbarus* and the Auditory Reaction Time. *Ibis* **107**, 101–104.

Grimes, L. 1966. Antiphonal singing and call notes of *Laniarius barbarus. Ibis* **108**, 122–126.

Hailman, J. P. 1967. The ontogeny of an instinct. The pecking response in chicks of the laughing gull *(Larus atricilla* L.) and related species. *Behaviour Suppl.* **15.**

Hinde, R. A. 1958. Alternative motor patterns in chaffinch song. *Anim. Behav.* **6**, 211–218.

Hjorth, I. 1970. A comment on graphic displays of bird sounds and analysis with a new device, the Melograph Mona. *J. Theor. Biol.* **26**, 1–10.

Holland, J. G., and Skinner, B. F. 1961. "The Analysis of Behavior." McGraw-Hill, New York.

Hooker, T., and Hooker, B. I. 1969. Duetting. *In* "Bird Vocalizations" (R. A. Hinde, ed.), pp. 185–205. Cambridge Univ. Press, London and New York.

Hutchison, R. E., Stevenson, J. G., and Thorpe, W. H. 1968. The basis for individual recognition by voice in the Sandwich tern *(Sterna sandvicensis). Behaviour* **32**, 150–157.

Impekoven, M. 1969. Auditory stimuli for pecking in the gull chick *Larus atricilla*. Effect of prenatal experience on postnatal behavior. Paper presented at the 11th International Ethological Conference, Rennes. Unpublished.

Kendeigh, S. C. 1945. Nesting behavior of wood warblers. *Wilson Bull.* **57**, 145–164.

Kimble, G. A. 1961. "Hilgard and Marquis' Conditioning and Learning. Appleton-Century-Crofts, New York.

Kirkman, F. B. 1937. "Bird Behaviour." Nelson, London.

Klopfer, P. H., and Gamble, J. 1966. Maternal "imprinting" in goats. The role of chemical senses. *Z. Tierpsychol.* **23**, 588–592.

Konishi, M., and Nottebohm, F. 1969. Experimental studies in the ontogeny of avian vocalizations. *In* "Bird Vocalizations" (R. A. Hinde, ed.), pp. 29–48. Cambridge Univ. Press, London and New York

Kruuk, H. 1964. Predators and anti-predator behaviour of the black-headed gull (*Larus ridibundus* L.). *Behaviour Suppl.* **11**.

Lanyon, W. E. 1963. Experiments on species discrimination in *Myiarchus* flycatchers. *Amer. Mus. Nov.* **2126**, 1–16.

Lanyon, W. E. 1967. Revision and probable evolution of the *Myiarchus* flycatchers of the West Indies. *Bull. Amer. Mus. Natur. Hist.* **136**, 331–370.

Lanyon, W. E. 1969. Vocal characters and avian systematics. *In* "Bird Vocalizations" (R. A. Hinde, ed.), pp. 291–310. Cambridge Univ. Press, London and New York.

Lashley, K. S. 1915. Notes on the nesting activities of the noddy and sooty terns. *Carnegie Inst. Wash.* **7**, 301–366.

Lorenz, K. 1931. Beiträge zur Ethologie sozialer Corviden. *J. Ornithol.* **79**, 67–127.

Lorenz, K. 1935. Der Kumpan in der Umwelt des Vogels. *J. Ornithol.* **83**, 137–413.

Lorenz, K. 1937. Über die Bildung des Instinktbegriffes. *Naturwissenschaften* **25**, 289–300, 307–318, 325–331.

Lorenz, K. 1938. A contribution to the comparative sociology of colonial-nesting birds. *Proc. 8th Int. Ornithol. Congr., Oxford, 1934* pp. 204–218.

Lorenz, K. 1950. The comparative method in studying innate behaviour patterns. *Symp. Soc. Exp. Biol.* **4**, 221–268.

Marler, P. 1967a. Comparative study of song development in sparrows. *Proc. 14th Int. Ornithol. Congr., Oxford, 1966* pp. 231–244.

Marler, P. 1967b. Animal communication signals. *Science* **157**, 769–774.

Marler, P. 1969. Tonal quality of bird sounds. *In* "Bird Vocalizations" (R. A. Hinde, ed.), pp. 5–18. Cambridge Univ. Press, London and New York.

Marler, P., and Hamilton, W. J. 1966. "Mechanisms of Animal Behavior." Wiley, New York.

Marler, P., and Isaac, D. 1960a. Physical analysis of a simple bird song as exemplified by the chipping sparrow. *Condor* **62**, 124–135.

Marler, P., and Isaac, D. 1960b. Song variation in a population of brown towhees. *Condor* **62**, 272–283.

Messmer, E., and Messmer, I. 1956. Die Entwicklung der Lautäusserungen und einiger Verhaltensweisen der Amsel (*Turdus merula merula* L.) unter natürlichen Bedingungen und nach Einzelaufzucht in schalldichten Räumen. *Z. Tierpsychol.* **13**, 341–441.

Moynihan, M. 1955. Some aspects of reproductive behaviour in the black-headed gull (*Larus r. ridibundus*) and related species. *Behaviour Suppl.* **4**.

Nice, M. M. 1943. Studies in the life history of the song sparrow. II. The behavior of the song sparrow and other passerines. *Trans. Linn. Soc. N.Y.* **6**, 1–328.

Nice, M. M. 1962. Development of behavior in precocial birds. *Trans. Linn. Soc. N.Y.* **8**, 1–211.

Nicolai, J. 1956. Zur Biologie und Ethologie des Gimpels (*Pyrrhula pyrrhula* L.). *Z. Tierpsychol.* **13**, 93–132.

Noble, G. K., and Wurm, M. 1943. The social behavior of the laughing gull. *Ann. N.Y. Acad. Sci.* **45**, 179–220.

Patterson, I. J. 1965. Timing and spacing of broods in the black-headed gull (*Larus ridibundus* L.). *Ibis* **107**, 433–460.

Penney, R. L. 1962. Breeding behavior is influenced by vocal recognition of voices of the Adelie. *Natur. Hist.* **71**, 16–25.

Pumphrey, R. J. 1961. Hearing in birds. *In* "Biology and Comparative Physiology of Birds" (A. J. Marshall, ed.), Vol. 2, pp. 55–68. Academic Press, New York.

Richdale, L. E. 1957. "A Population Study of Penguins." Oxford Univ. Press, London and New York.

Rittinghaus, H. 1953. Adoptionsversuche mit Sand- und Seeregenpfeifern. *J. Ornthol.* **94**, 144–159.

Robinson, A. 1946. Magpie-larks: A study in behaviour. *Emu* **46**, 265–281, 382–391.

Robinson, A. 1947. Magpie-larks: A study in behaviour. *Emu* **47**, 11–28, 147–153.

Schwartzkopf, J. 1962. Vergleichende Physiologie das Gehors und der Lautausserungen. *Fortschr. Zool.* **15**, 214–336.

Siegel, S. 1956. "Non-Parametric Statistics." McGraw-Hill, New York.

Sladen, W. J. L. 1958. The Pygoscelid penguins. I. Methods of study. II. The Adelie penguin *Pygoscelis adeliae* (Hombron and Jacquinot). *Falkland Isl. Depend. Surv., Sci. Rep.* **17**.

Smith, W. J. 1968. Message-meaning analysis. *In* "Animal Communication" (T. A. Sebeck, ed.), pp. 44–60. Indiana Univ. Press, Bloomington, Indiana.

Stein, R. C. 1956. A comparative study of advertising song in the Hylocichla thrushes. *Auk* **73**, 503–512.

Stewart, R. E. 1953. A life history study of the yellowthroat. *Wilson Bull.* **65**, 99–115.

Stonehouse, B. 1960. The king penguin *Aptenodytes pategonica* of South Georgia. I. Breeding behaviour and development. *Falkland Isl. Depend. Surv., Sci. Rep.* **23**.

Stout, J. F., Wilcox, C. R., and Creitz, L. E. 1969. Aggressive communication by *Larus glaucescens*. Part 1. Sound communication. *Behaviour* **34**, 29–41.

Thielcke, G. 1962. Versuche mit Klangattrappen zur Klärung der Verwandschaft der Baumläufer *Certhia familiaris* L., *C. brachydactyla* Brehm und *C. americana* Bonaparte. *J. Ornithol.* **103**, 266–271.

Thielcke-Poltz, H., and Thielcke, G. 1960. Akustisches Lernen verscheiden alter schallisolierter Amseln *Turdus merula* L. und die Entwicklung erlernter Motive ohn und mit Künstlichem Einfluss von Testosteron. *Z. Tierpsychol.* **17**, 211–244.

Thompson, D. H., and Emlen, J. T. 1969. Parent-chick individual recognition in the Adelie penguin. *Antarct. J.* **3**, p. 132.

Thompson, N. S. 1969. Individual identification and temporal patterning in the cawing of common crows. *Commun. Behav. Biol.* Commun. No. 09690044.

Thorpe, W. H. 1958. The learning of song patterns by birds, with especial reference to the song of the chaffinch, *Fringilla coelebs*. *Ibis* **100**, 535–570.

Thorpe, W. H. 1961. "Bird Song. The Biology of Vocal Communication and Expression in Birds." Cambridge Univ. Press, London and New York.

Thorpe, W. H. 1963. Antiphonal singing in birds as evidence for avian auditory reaction time. *Nature (London)* **197**, 774–776.

Thorpe, W. H. 1967. Vocal imitation and antiphonal song and its implications. *Proc. 14th Int. Ornithol. Congr., Oxford, 1966* pp. 245–263.

Thorpe, W. H. 1968. Perceptual basis for group organization in social vertebrates, especially birds. *Nature (London)* **220**, 124–128.

Thorpe, W. H., and North, M. E. W. 1966. Vocal imitation in the Tropical Bou-bou shrike *laniarius aethiopicus major* as a means of establishing and maintaining social bonds. *Ibis* **108**, 432–435.

Tinbergen, N. 1951. "The Study of Instinct." Oxford Univ. Press, London and New York.

Tinbergen, N. 1953. "The Herring Gull's World." Collins, London.

Tinbergen, N. 1959. Comparative studies of the behaviour of gulls (Laridae): a progress report. *Behaviour* **15**, 1–70.

Tinbergen, N., Impekoven, M., and Frank, D. 1967. An experiment on spacing-out as a defence against predation. *Behaviour* **28**, 307–321.

Tschanz, B. 1965. Beobachtungen und Experimente zur Entstehung der "persönlichen" Beziehund zwischen Jungvogel und Eltern bei Trottellummen. *Verh. Schweiz. Naturforsch. Ges.* **1964**, 211–216.

Tschanz, B. 1968. Trottellummen. *Z. Tierpsychol. Suppl.* **4.**

Watson, J. B., and Lashley, K. S. 1915. Homing and related activities of birds. *Carnegie Inst. Wash. Publ.* **211.**

Watson, M. 1969. Significance of antiphonal song in the Eastern whipbird, *Psophodes olivaceus. Behaviour* **35**, 157–178.

Weeden, J. S., and Falls, J. B. 1959. Differential responses of male ovenbirds to recorded songs of neighbouring and more distant individuals. *Auk* **76**, 343–351.

White, S. J., and White, R. E. C. 1970. Individual voice production in gannets. *Behaviour,* in press.

White, S. J., White, R. E. C., and Thorpe, W. H. 1970. The acoustic basis for individual recognition by voice in the gannet. *Nature (London)* **225**, 1156–1158.

Ontogenetic and Phylogenetic Functions of the Parent-Offspring Relationship in Mammals

LAWRENCE V. HARPER[1]

CHILD RESEARCH BRANCH, NATIONAL INSTITUTE OF MENTAL HEALTH
BETHESDA, MARYLAND

I. Introduction

The analysis of parent-offspring relations in mammals has been the object of considerable attention (e.g., Rheingold, 1963b; Richards, 1967). However, despite well-documented arguments for considering the functional effects of caretaking as symmetrical, affecting both young and parent (e.g., Lehrman, 1956; Rosenblatt *et al.*, 1961; Schneirla and Rosenblatt, 1961), stress has been laid upon the effects of caretaking as they are manifest in the behavior of the young. Ontogenetically, the parental role (apart from facilitating physiological development) has been described in terms of shaping the responses of the offspring toward their physical and social environments (e.g., Rheingold, 1963a). Phylogenetically, the adaptive value of parental care in mammals has been attributed to an increased potential for adjustment to the environment due to caretaker tuition (e.g., Romer, 1967). In both cases the emphasis is on the modification ("socialization") of offspring behavior.

[1]Present address: Department of Applied Behavioral Sciences, University of California, Davis, California.

75

The purpose of this review is to summarize and evaluate available data that are relevant to these views and to marshall a substantial body of evidence which indicates that (1) the influence of caretakers on offspring behavioral development may have been overestimated, (2) the caretaker may be comparably affected by the young, and (3) due to the facilitating effects of the parent-offspring relationship, the young of some forms may play a major role in the adaptive radiation of their species. The paper is divided into three major sections dealing with each of the above points in turn.

II. SOCIALIZATION: THE EFFECTS OF THE PARENT-OFFSPRING RELATIONSHIP ON THE YOUNG

Since each species is characterized by a unique complex of behavioral adjustments to its surroundings (cf. Tinbergen, 1960), the requirements of the young and the means employed by caretakers to meet them vary among mammals. In some mammals, the nature of the species' adaptation suggests that the young must be capable of acquiring certain responses in the absence of parental tuition. In others, however, there seems to be ample opportunity for and benefits to be obtained from caretaker guidance. Negative findings in the latter species would have considerable theoretical significance since the presence of conditions propitious for learning often is taken as presumptive evidence for its occurrence.

In 1962, after a review of the literature, Collias concluded that "guidance" by the parents and older siblings widened the young's diet and geographical range, conditioned intraspecific cooperation, and led to the avoidance of predators (Collias, 1962). He described the caretaker-offspring relationship as affecting the formation of social bonds between the young and their conspecifics, the differentiation of the offspring's social responses, and the orientation of the young in the physical and social environments, including their integration into new groups. Since that time, evidence for the effects of caretaking on young mammals' ecological adaptations and social relationships has continued to accumulate.

A. RELATING THE OFFSPRING TO THE PHYSICAL ENVIRONMENT

The location of the birth site determines the offspring's first contacts with their inanimate environment; subsequent widening of environmental contacts may be subject to regulation by the parent.

1. Regulating Movement in Space

Field and observational studies indicate that caretakers often attempt to restrain the investigatory behavior of the young (Hafez *et al.*, 1962; Hinde and Spencer-Booth, 1967a; Kaufmann, 1966; Menzel, 1966, 1967), and laboratory experiments show that the mother's "emotionality" affects the later exploratory behavior of her offspring (Denenberg, 1967). However, the only experimental study of social transmission of avoidance responses (Stephenson, 1967) known to the writer yielded somewhat mixed results. Three-year-old rhesus monkeys, *Macaca mulatta*, were divided into like-sexed pairs; each subject was given individual avoidance training to one of two objects while its partner received similar training to the other object. When stable avoidance responses had been obtained, the animals were tested together. In three of four male pairs, the "naive" animal's manipulation of a test object was reduced from pretraining control levels when it was tested with the "experienced" partner—despite the fact that no unconditioned stimulus was presented. However, in three of four female pairs, the trained animals displayed more manipulation than they had in their individual post-avoidance-training trials. Thus, only the possibility, not the regularity, of social transmission of avoidance behavior was demonstrated.

While the influence of conspecific restraint remains questionable, there exists good evidence that interaction with adult congeners is not necessary for the display of exploratory or manipulative responses. The presence of inanimate, terry-cloth mother surrogates in otherwise frightening settings was sufficient to facilitate investigatory behavior in isolation-reared rhesus monkeys (Harlow, 1961, 1962), and, even in the absence of any parent-object, isolates displayed an increasing interest in test objects during their first three months (Mason *et al.*, 1959; see Butler, 1965, for a review).

2. Homesite and Habitat

It has been suggested that parents may play an important role in acquainting their offspring with their home ranges (Schenkel, 1966b) and preparing them to deal with problems posed by the terrain (Altmann, 1963). The tendency of some chiroptera (e.g., Cockrum, 1956) and pinnipeds (e.g., Batholomew and Hoel, 1953; Scheffer, 1950; Smith, 1968) to congregate in "traditional" breeding grounds implies that early experience in the caretaking situation may affect later movements or habitat preferences. As adults, Alaska fur seals, *Callorhinus ursinus,* return to their birthplaces to mate and bear their young (Scheffer, 1950), and an experiment with sheep, *Ovis aries,* by Hunter and Davies (1963) suggests the existence of a sensitive period for an ewe's becoming attached to a grazing site.

However, in the prairie deermouse, *Peromyscus maniculatus bairdii,* there is evidence that prior experience during development constitutes only one of at least two mechanisms underlying later habitat preference. Wecker (1963) reared the offspring of wild-trapped and laboratory-bred *P. m. bairdii* in a laboratory colony, in outdoor pens in the open, or in outdoor pens in a wooded area. When adult, the animals from all three groups were tested for their choice of nest sites in another outdoor enclosure, half of which was in a wooded area, the other half in the open. He found that, regardless of the environment in which they had been reared, the offspring of the wild-trapped animals spent more time in the open portion of the test enclosure. In contrast, the offspring of animals from the 12th to the 20th generations of laboratory-bred *P. m. bairdii* showed no such preference in the test field if they had been reared in either the wooded enclosure or the laboratory. However, the open field-reared offspring of laboratory-bred mice built their nests in the open half of the field. These data led Wecker to suggest that habitat preference in this species may depend upon two factors: an experientially developed preference for an open environment, and a less strongly buffered tendency to select a nest site in low grassland regardless of prior experience.

3. Feeding Behavior

A number of observations of wild and captive animals suggest that the eating habits of the young may be influenced by association with caretakers. Food preferences may be affected indirectly, where the mother's intake alters the taste of her milk (LeMangen and Tallon, 1968) or directly, where the young actually take food from their caretakers (e.g., Booth, 1962; Epple, 1968; Ewer, 1963; Ruffer, 1965; Schaller, 1963; Tevis, 1950; van Lawick-Goodall, 1967). In some carnivores, the mother has been reported to facilitate her offspring's enactment of certain aspects of prey-killing behavior (Ewer, 1969; Leyhausen, 1968). In Japanese macaques, *Macaca fuscata,* there is even evidence of troop-specific traditional feeding habits being transmitted from caretaker to offspring (Kawamura, 1959).

However, in some mammals, species-typical feeding or predatory patterns may appear without tutelage. Red squirrels, *Sciurus vulgaris,* were taken from their mothers before their eyes opened; until they were

60–75 days old, they were allowed no opportunity to dig or to manipulate solid objects. When presented with hazel nuts at 75 days, they went through the full sequence of acts involved in nut-burying—even where there was no earth to dig in. Although inefficient with respect to the characteristics of the nutshell, their nut-gnawing and spitting movements were species-typical (Eibl-Eibesfeldt, 1967). Domestic cats, *Felis catus*, were reared in individual wiremesh cages from the age of 5 days. At one year of age, the isolates' attack behavior toward a rat was elicited by hypothalamic stimulation. Compared to attacks similarly elicited in socially reared animals, the isolates' responses were somewhat less specific and intense but were organized normally and included all the components of the normal pattern. It was concluded that the development of the isolates' attack response and its neural organization was not dependent upon experience with prey or other cats (Roberts and Bergquist, 1968). Similarly, polecats, *Putorius putorius* L., which were hand-reared from the age of 21 days attacked and killed rats in the same way as did animals which were left with their littermates (Eibl-Eibesfeldt, 1961).

Thus, even in species in which parental behavior facilitates certain aspects of feeding behavior, those patterns may develop in the absence of such experience.

4. Dispersal

Frequently the young (particularly males) are driven from the parental homesite as they reach reproductive maturity (cf. Jewell and Loizos, 1966; Wynne-Edwards, 1962). Aside from its effect on regulating population density and the impetus it provides for expansion of the species' territory, this enforced dispersal has often been interpreted as an aspect of the parental educative process. However, as mentioned above, the young of many species apparently tend to explore their surroundings "on their own," and as immatures, their wanderings may cause them to violate the territory of other groups (e.g., Scott and Fuller, 1965; Koford, 1957). To the writer's knowledge, there is no evidence that enforced dispersal is a necessary condition for the colonization of unexploited habitats or for the development of species-typical behaviors.

B. Relating the Offspring to the Animate Environment

1. Predators

During the period of dependence upon the parent, mammalian young are afforded protection against predators by a wide variety of behavior patterns ranging from active combat with intruders (e.g., wolves, *Canis lupus:* Murie, 1944) to a combination of concealment and infrequent contact (e.g., European rabbits, *Oryctolagus:* Zarrow *et al.,* 1965). In those species where the young are subject to predation while in the company of caretakers, it has been suggested that the association of parental flight (Kabet *et al.,* 1953; also cited in Collias, 1962; Washburn and Hamburg, 1965), alarm calls (Booth, 1962), or mobbing responses (Epple, 1968; Jolly, 1966b) with the presence of potential predators may provide the means by which the young learn to avoid danger. However, no direct, experimental evidence demonstrating the necessity for such learning is known to the writer.

In contrast, there are data which indicate that, in some species, caretaker tutelage is not necessary for the development of flight reactions to humans. Gilbert (1968) observed that the fawn of a hand-reared fallow deer, *Dama dama,* which had had no contact with man from birth until at least the second day postpartum, avoided humans from the first encounter. Indeed, despite the fact that the young of this species normally remain close to their mothers, the fawn kept its distance even while the mother was allowing herself to be handled by humans. The young of zoo-reared wolves (Woolpy and Ginsburg, 1967) and domestic dogs, *Canis familiaris* (Scott and Fuller, 1965) displayed avoidance responses to humans unless they had been "socialized" to man in the first few weeks of life—regardless of parental responses. Even prematurely (cesarean) delivered guinea pigs, *Cavia porcellus,* which had been left alone with a lactating female for as little as 10 hours responded to humans with flight behavior (Avery, 1928).

Thus, it is clear that, in some mammals, avoidance of potential predators may not depend upon prior experience with the alien species. (This problem is considered further in the following section.)

2. Conspecifics

a. Gregariousness. The parent-offspring relationship is also generally accepted to be an important source of intraspecific gregariousness. (For recent reviews and hypotheses concerning the processes involved, see Cairns, 1966; King, 1966; Salzen, 1967; Scott, 1962, 1967.)

A number of experiments indicate that social preferences can be manipulated by altering the stimulus qualities of the natural parent (Mainardi *et al.,* 1965; Marr and Gardner, 1965) and by cross-species fostering at birth (Denenberg *et al.,* 1964; Hersher *et al.,* 1963; Quadagno and Banks, 1968) or at weaning (Denenberg *et al.,* 1964; Nagy, 1965). Similarly, there is evidence that hand-reared animals may direct their social responses to human caretakers (e.g., Hediger, 1955; Kunkel and Kunkel, 1964; Jolly, 1966a,b; Sackett *et al.*, 1965; Sorenson and Conaway, 1966).

Atypical responses to conspecifics by young which have been isolated from social contact (e.g., Schneirla *et al.*, 1963; Harlow and Harlow, 1966) are often interpreted as evidence for the role of caretaker-offspring interaction in the development of social responsiveness. However, there are grounds for doubting the validity of such conclusions. Many data indicate that parent-offspring interactions are not necessary for the development of intraspecific gregariousness. Domestic kittens which were caged singly from an early age with a rat would mew and move about restlessly upon removal of their cagemate and quiet down at the latter's return (Kuo, 1930). However, if pairs of kittens were reared with rats, they showed such behavior only when their kitten cagemate was removed (Kuo, 1938). These data imply that, if given a choice, kittens would form social attachments to their own kind despite the absence of a parent. (It is not obvious from current conceptions of parental tuition why these kittens chose another kitten rather than a rat as a social companion.)

In at least two rodent species, early social interaction with congeners is not a necessary condition for intraspecific approach or tolerance. Both male and female domestic guinea pigs which had been hand-reared in individual opaque-walled cages from birth to the age of 80 days made as many approach responses to a like-sexed congener as did mother-reared animals (Harper, 1968a). Female rats, *Rattus norvegicus,* which had been incubator-reared from birth to reproductive maturity without any contact with other rats mated with adult mates and were "maternally" responsive toward their offspring (Thoman and Arnold, 1968a).

On the other hand, cross-fostering and hand-rearing experiments have clearly demonstrated that early experience in the caretaking setting can affect later gregariousness. The solution to this apparent contradiction is suggested by an experiment (Mainardi, 1963) in which female mice, *Mus musculus domesticus,* reared by their mothers alone showed no mating preference from same-strain males while those reared with their sires mated less often with "strange" *(M. m. bacitrans)* males. These data imply that in mammals, as in avians, early experience with the caretaker acts to delimit the range of otherwise adequate stimuli which will serve to elicit social or approach responses (Salzen, 1967; Sluckin, 1965).

Evidence consistent with this view is provided by a study (Harper, 1968b) in which guinea pigs were isolated at birth and for their first 3 days were allowed nine 20-minute exposures to a moving, calling model resembling a guinea pig in the nursing posture. At 9 days of age they were given a choice between their mother or the model in a large T-maze. They made significantly more model choices than did their mother-exposed or empty runway-exposed littermates. Yet, when the animals were returned to their mothers at the age of 10 days, all of them made satisfactory nursing adjustments. In a later experiment, a group of isolated guinea pigs was similarly exposed to a model during their first 3 days and then returned to their mothers; all were observed to make satisfactory nursing adjustments. After 7 days of social interaction with conspecifics, those pups which had displayed approach and filial responses to the model in their first 3 days of life were retested; they all responded socially to the model. In contrast, mother-reared pups, which had received comparable amounts of exposure to the model during their first 3 days of life or were exposed at 10 days of age made significantly fewer approach or "social" overtures to models. The data suggest that the early experience with a model acted to increase the likelihood of the animals' making approach responses to it, while more complete social isolation, or prior experience with a "more adequate" object, led to the tendency to avoid unfamiliar, moving objects. (Recall Kuo's kittens, p. 81).

The failure of some isolation-reared animals to become integrated into existing conspecific social groups cannot always be interpreted as being due to an absence of gregarious tendencies or to an avoidance of the unfamiliar. Four hand-reared fallow deer were introduced to a semi-wild herd at the age of about 9 months. Upon their approach to the adults, the fawns were met with bites and charges from the does (Gilbert, 1968). Thus, in other forms such as sheep or goats, *Capra,* in which the approach response of alien offspring are also rejected by adult females, the failure of hand-reared young to mix with an existing group may represent little more than the suppression of offspring approach behavior as a result of adult hostility.

Cross-species fostering, while indicating areas of behavioral malleability, also provides evidence of the resistance of certain response patterns to social influence. For instance, in experiments with sheep and goats (Hersher *et al.,* 1963), it was found that the kids reared by sheep and the lambs reared by goats retained their respective species-typical response tendencies: the lambs followed, and the kids left the mother and lay down, after suckling. These patterns persisted for the normal duration despite the repeated attempts of the ewes to retrieve their foster kids and the does to leave their resting lambs.

b. Peer Relationships. The widening of social interactions, as indicated by peer-directed behavior (play), may also be affected by the presence or behavior of the parent. Field studies indicate that the young of some species do not play in the absence of the mother (Linsdale and Tomich, 1953; Schenkel, 1966a), and laboratory experiments with macaque monkeys show that separation from the mother leads to a reduction of play activity, whether within an otherwise familiar social context (Hinde *et al.*, 1966; Kaufman and Rosenblum, 1967; Spencer-Booth and Hinde, 1967) or in a novel setting (Alexander and Dodsworth, 1966; Seay *et al.*, 1962; Seay and Harlow, 1965). Early social isolation (Mitchell, 1968) or peer contact in the absence of the mother (Harlow and Harlow, 1965) reduces or slows, respectively, the development of play. Further, the play of mother-reared animals reflects the quality of mothering behavior (Arling and Harlow, 1967; Mitchell *et al.*, 1966).

These data indicate that parental presence or caretaking practices may affect the display of peer-directed behavior in several species of mammals. However, there is no evidence which indicates that experience with a caretaker is a precondition either for attraction toward peers or for the development of play.

The tendency to direct affiliative behavior toward age-mates has been observed in mother-deprived mammals. The finding that kittens which were reared apart from their mothers displayed "attachment responses" to their littermates rather than to an adult rat cagemate (Kuo, 1930) has already been noted. Gilbert's (1968) hand-reared fallow deer fawns grouped together upon their being rejected by the adult does in the herd. The presence of a caretaker did not appear to be required for the display of approach behavior toward conspecific age-mates in these species.

In domestic guinea pigs, there is evidence that the tendency to respond positively toward peers may be present even before term. Avery (1928) reported that fetuses delivered by cesarean section up to 4 days prematurely would move to one another and protest vocally if separated. The writer (unpublished) observed that hand-reared infant guinea pigs which had been exposed to moving models would approach and display what appeared to be play responses toward one another upon their first opportunity to interact with a peer. In domestic dogs, Fuller and Clark (1966b) reported that pairs of beagle puppies which were reared together from the age of 3 weeks in barren, opaque-walled cages displayed normal peer-directed social interactions while in the home cage. Similar data have been reported in primates. Alexander and Harlow

(1965) separated four rhesus monkeys from their mother at birth and reared them together in a single cage. The animals displayed the full range of age-appropriate peer directed behavior observed in mother-reared animals, albeit at a somewhat reduced intensity. Subsequent variations of this experiment have yielded comparable results (Harlow and Harlow, 1969).

Thus, there is evidence that, in several mammalian species, neither the activity nor the presence of a caretaker is necessary for the development and display of peer directed social behavior.

c. Social Rank. Various experiences in the caretaking context have been cited as possible precursors of adult dominance. Field studies suggest that caretakers may provide "models" for the later assertive behavior of the young (Calhoun, 1962; Rabb *et al.*, 1967; Sade, 1967). However, in these studies, the effects of modeling are confounded with the possibility that assertive tendencies are inherited and the fact that the young of more dominant animals may be subject to fewer debilitating stresses arising from social competition. Likewise, the currently available experimental data are not without alternative interpretations. The deficiencies observed in hand-reared animals may be the result of rejection of aliens by the group (e.g., Gilbert, 1968; Masserman *et al.*, 1968) or, in the case of isolates, low dominance may result from the behavioral disorganization characteristic of the "postisolation syndrome" (Fuller and Clark, 1966a,b; also see below). Thus, there exist no unequivocal demonstrations that parental guidance accounts for the later social rank of adult mammals.

d. Aggression. Several experiments have shown that caretaking experiences affect later aggressive behavior. Fostering infant male mice on rat mothers reduces the frequency of intraspecific fighting in the mice at adulthood (Denenberg *et al.*, 1964). The amount of maternal rejection or punishment of their offspring by rhesus monkey mothers can be related to the level of aggression shown by the young as adolescents (Arling, 1966; Sackett, 1967) or juveniles (Mitchell *et al.*, 1967).

These studies demonstrate that the level of intraspecific hostility may be affected by experience in the caretaking setting, but they do not demonstrate that caretaker tuition is a necessary condition for the display of such behavior. Indeed, uncontrolled aggressive outbursts have been described as one of the sequelae of isolation in rhesus monkeys (Harlow and Harlow, 1965).

e. Communication. In pigtailed macaques, *Macaca nemestrina*, there is evidence of group-specific vocal "dialects" (Grimm, 1967), and isolation-reared rhesus monkeys fail to respond to the postural and other signals emitted by their peers (Harlow and Harlow, 1965; Mason, 1963; Miller *et al.*, 1967). However, only Grimm's findings suggests the social transmission of communicative behavior patterns. (The isolation studies pose certain problems for interpretation which will be discussed in greater detail in Section II,C.) Further, there are findings which suggest that the performance of and response to certain vocal signals or postural displays may occur appropriately without prior social experience. Booth (1962) reported that hand-reared brazza, *Cercopithecus neglectus,* and vervet, *C. aethiops,* monkeys took flight when first hearing the alarm call of their species. The writer observed that a hand-reared male guinea pig, upon his first exposure to another male of his species, took flight and responded in kind when the latter began tooth-chattering (a threat display). Sacket (1966) reported that isolation-reared rhesus monkeys which were exposed to the projected photographic images of various objects, including a conspecific in the threat posture, responded to the latter with agitated behavior at about that age at which rhesus mothers begin to direct threat responses toward their offspring (cf. Hansen, 1966).

Thus, the development of communicative behavior in some mammals may occur in the absence of parental educative influences, while in others, tradition may account for variations in some, but not necessarily all, aspects of the group repertoire.

f. Reproductive Behavior. The results of social isolation are consistent with the view that early caretaking behavior may affect the later reproductive performance of the young. Male laboratory rats which were isolated from birth (Gruendel and Arnold, 1969) or at weaning (Folman and Drori, 1965; Gerall *et al.,* 1967; Hård and Larsson, 1968; Zimbardo, 1958) displayed fewer successful copulations than did socially reared animals. Similar findings have been reported in male guinea pigs (A. A. Gerall, 1963; H. D. Gerall, 1965; Young, 1957), dogs (Beach, 1968), and cats (Rosenblatt and Schneirla, 1962). In chimpanzees, *Pan troglodytes* (Nissen, 1956; cited by Young, 1957), and rhesus monkeys (Harlow and Harlow, 1966; Mason, 1963), the copulatory responses of both males and females were disturbed as a result of social isolation. Aberrant maternal behavior has been observed in isolation-reared rhesus monkeys (Seay *et al.,* 1964; Arling and Harlow, 1967) and zoo-reared chimpanzees (Hediger, 1965).

Although early isolation may lead to disturbed reproductive behavior in mammals, it is another matter to conclude that social interaction is necessary for the development of species-typical behavior patterns. For example, Beach found that male rats which were reared in isolation from the age of 14 (Beach, 1958) or 21 days (Beach, 1942) mated as frequently as did socially reared controls. Since 14-day-old rats are barely capable of locomotion, it is quite unlikely that adult copulatory patterns could have been shaped by parental guidance during the nursing period. Similarly, in guinea pigs, Gerall (1963) showed that exposure to conspecifics through a wide-mesh wire screen was sufficient for the development of normal copulatory behavior in males. Further, the writer observed that despite their having been isolated from birth to the age of 80 days, two of five male guinea pigs made normally oriented mounts accompanied with pelvic thrusts on their first exposure to a receptive female (Harper, 1968a). Even in rhesus monkeys, there is evidence that actual contact or interactions with a species mate may not be a prerequisite for the ontogenetic organization and display of coitus-related acts. Mason (1968) found that infant rhesus monkeys which were reared from birth with cloth covered "artificial mothers" that moved freely on an irregular schedule displayed age-appropriate presenting, and rudimentary mounting responses to the surrogate.

Similar considerations obtain for parental behavior. Thoman and Arnold (1968a) found that female laboratory rats which had been reared from birth in an incubator, with or without peer contact, until a brief encounter with a potent male at about 85 days of age showed normal parturition, nest building, and retrieving behavior upon littering. In rhesus macaques, there is evidence that rearing females alone from birth with only auditory and visual contact with other monkeys may not, in itself, suffice to impair grossly the caretaking efficiency of primipara. Meier (1965) reported that the behavior of females so reared became inadequate or rejecting only in cases in which the animals' infants were delivered by cesarean section and not presented to the mothers for some (unspecified) period after delivery. In contrast to zoo-reared chimpanzees, Lemmon (1968) reported that, despite no prior caretaking experience or opportunities to observe caretaking, two of four primiparous, laboratory-reared chimpanzees delivered and cared for their offspring in a normal manner.

Thus, the development of reproductive behavior patterns in some mammals may not depend upon prior social interaction. Further, the data suggest that in the interpretation of isolation experiments, the failure to perform, or distortion of, behavior patterns as a result of restricted experience do not necessarily imply that exposure to exogenous influences is a prerequisite for the organization of these patterns. Often, the conditions of testing may block the expression of a normal potential (cf. Eibl-Eibesfeldt, 1961, and below).

C. The Effects of Isolation

Since isolation experiments are often cited in support of the view that social experience is required for the development of intraspecific responses, it will be helpful to examine the effects of such treatment.

1. The Display of Species-Typical Behavior Patterns

There is increasing evidence that isolation-induced behavioral aberrations may represent a situational disruption of species typical behavior, rather than the unavailability of such response patterns. Zimbardo (1958) suggested that isolated rats were more reactive to environmental disturbances. The possible dynamics of this effect have been suggested by Fuller and his collaborators (for a review, see Fuller, 1967), who reared domestic puppies from their third week to various ages in a variety of highly-controlled surroundings isolated from human contact. Relative to laboratory reared controls, isolation from any contact with peers or human handlers until the age of 15 weeks produced severely aberrant behavior. However, puppies reared in individual cages without human handling, but allowed to look out a window into the laboratory were less disturbed than total isolates, and single weekly, 20-minute exposures in the test situation were sufficient to completely overcome the effects of isolation. Furthermore, administration of a tranquilizer (chlorpromazine) before handling and before and after testing decreased the amount of deficit in total-isolation-reared animals. Of special interest was the fact that a few subjects showed no ill effects from experiential deprivation, either in learning tests or in the arena situation.

On the basis of these findings, Fuller concluded that the explanation most likely to account for isolation-induced deficits in the behavior of puppies was that the isolated animals became habituated to a very low level of stimulation; when they were suddenly thrust into a novel, complex (test) situation, they became hyperexcitable and unable to respond selectively to stimuli. Electroencephalographic studies of free-moving, isolation-reared dogs support this view of an excessive "arousal" response in complex settings (Melzack and Burns, 1965; see also Melzack, 1968).

Comparable findings have been obtained with primates. Investigations of "abnormal" responses in isolated infant chimpanzees (Davenport and Menzel, 1963) and crab-eating macaques, *Macaca fascicularis,* (Berkson, 1968) have shown that a change in the environment or the presence of novel objects tended to elicit such behavior. Berkson (1968) reported that the behavior of undisturbed isolate crab-eating monkeys in their home cages often was difficult to distinguish from that of socially reared animals; abnormalities occurred only in response to environmental disturbances.

Mitchell (1968) tested the social responsiveness of 3½- to 4½ -year-old rhesus monkeys which had been reared in isolation from birth to the age of 6 months, from birth to 12 months, or from 6 to 12 months of age. Compared to age-mates which had been reared by their mothers and had experience with peers, the isolates exhibited more fear grimaces and attempted flight and, especially the females, more aggression toward infants. All the isolates spent more time immobile than did the socially experienced controls. However, the older isolates generally appeared to be "less disturbed." Mitchell suggested that the isolates' responses might be "rehabilitated" by exposure to infants since the latter elicited the least "fear."

There is some evidence in primates that experiential deprivation may not affect the development (as opposed to the display) of most motor patterns. Isolation affected the inclination to perform or the organization of juvenile "sexual" patterns of rhesus monkey infants, not the capability for performing the elements of the pattern (Alexander and Harlow, 1965). Even Mason's (1963) isolation-reared rhesus monkeys displayed all the elements of the normal coital pattern, although their overall behavior was aberrant. (The latter observations have been replicated by Senko, 1966). Menzel (1964) separated infant chimpanzees from their mothers on their first day of life and reared them in various degrees of isolation for 21 months. Although the isolates displayed more variable patterns of response and "autistic" behavior, by the age of 4 or 5 years they had performed every discrete act observed in the socially reared controls. The evocation of normal behavior seemed to depend upon a reduction of excessive "arousal" rather than a specific learning situation.

2. Stereotyped Behavior

Isolation experiments show that, both in chimpanzees (Davenport *et al.*, 1966) and crab-eating macaques (Berkson, 1968), the later the separation from the mother, the less marked the incidence of stereotypy. Further, by comparing the behavioral development of mother-reared infants with that of infants which were isolated at birth or after 1, 2, 4, or 6 months with their mothers, Berkson was able to demonstrate that a number of apparently bizarre, repetitive acts could reasonably be interpreted as either intensifications of responses shown transitorily by mother-reared animals or displacements of infantile behaviors in the absence of appropriate objects—i.e., lacking suitable objects for the normal expression of more mature patterns, these motor acts persisted and interfered with the expression of the latter.

In rhesus monkeys, Mason (1968) has shown that, while infants reared in isolation with inanimate, terrycloth-covered mother-surrogates displayed stereotyped rocking and self-clasping behavior, isolates reared with identical surrogates which moved intermittently did not. Indeed, at later ages, the animated-surrogate-reared infants showed a variety of age-appropriate, species-typical behaviors. Thus, in rhesus, neither the presence of conspecifics nor contingent responses from mother-surrogates are required for development which more closely approximates the species-typical pattern.

D. SUMMARY

The data currently available lead to the conclusion that the role of the caretaker in the behavioral development of mammalian young may be considerably more circumscribed than has been heretofore believed. In particular, the influence of the parent on the development of species-typical social responses has been overestimated. It is true that in some species, the transmission of social traditions, including feeding patterns and the selection of homesites, may be largely determined by caretakers. However, aside from the provision of food, shelter, and physical protection, the most commonly observed contributions of the mammalian parent seem to be the provision of a stable context within which the behavioral development of the young may take place, and, perhaps, a modulatory influence on the intensity or extensity of the display of such behavior. The aberrations associated with isolation experiments, especially where appropriately performed "elements" occur, may often represent an inability to respond selectively to those details of the test environment which are most salient to animals which have previously had an opportunity to become familiar with complex stimuli—not a failure to learn the responses themselves (see also Bronson, 1968; Fuller, 1967; Jensen and Bobbitt, 1967).

Thus, it may be profitable to extend the search for the evolutionary significance of caretaking in mammals. Schneirla and his co-workers (Schneirla et al., 1963; Rosenblatt et al., 1961) have emphasized the importance of the mutual stimulation that occurs during the parent-offspring relationship within a caretaking cycle. The existence of such reciprocity raises the possibility that an examination of the effects of the young on their caretakers may help to elucidate the role of parental behavior in mammalian evolution. In the following section, evidence will be presented which indicates that for each general way in which caretakers affect the behavior of the young, the offspring of one or more species exert a similar effect on their parents.

III. EFFECTS OF THE PARENT-OFFSPRING RELATIONSHIP ON THE PARENTS

Just as the degree of caretaker involvement (and caretaker influence on the behavior of the offspring) varies across species, so does the magnitude of the effects that the young exert upon their parents. Unfortunately, the data are insufficient to permit an evaluation of the degree of reciprocity or to relate it to the extensity of parental solicitude required. However, evidence for the existence of such effects is substantial.

A. RELATIONSHIP OF THE PARENT TO THE PHYSICAL ENVIRONMENT

The caretaker's relationship to its physical surroundings is often affected by the presence of the young.

1. Regulating Movement in Space

While the parent may play a decisive role in orienting its offspring in space, the relationship may be reciprocal. In some precocial animals, the location of resting or hiding sites is often determined by the young (Collias, 1956; Estes, 1967; Frädrich, 1965; Linsdale and Tomich, 1953) or the area around the moving offspring is the focus of parental activity (Altmann, 1958; Blauvelt, 1956). In those mammals which rear their offspring in nests or dens, such sites likewise become the hub of caretaker activity in the early postpartum period (e.g., Bleicher, 1962; Eisenberg, 1963; Ewer, 1967; McCarley, 1966; Murie, 1944).

Parental activity may vary according to the stage of the caretaking cycle. Mother Dall sheep, *Ovis dalli,* in Alaska congregated in the more rugged areas of their range while their lambs were young (Murie, 1944). Female chipmunks, *Eutamis,* moved about less and had smaller ranges than did males and nonpregnant females. Upon emergence of their offspring, the mothers' ranges increased or they spent more time above ground (Martinsen, 1968). Similarly, the movements of a free-ranging pack of Canadian wolves were shorter when the pups were young (Joslin, 1967). The latter two examples suggest that the motor capacities or proclivities of the young may play a part in limiting caretaker movement. Field studies indicate that the reluctance to swim of very young caribou calves, *Rangifer tarandus groenlandicus* L., may reduce the mobility of the entire band (de Vos, 1960; Lent, 1966), and the burden imposed by the necessity to carry the young may reduce the mobility of gibbons, *Hylobates lar* (Carpenter, 1940), and Hamadryas baboons, *Papio hamadryas* (Kummer, 1968).

2. Homesite and Habitat

In some mammals, caretaking may be accompanied by a change in a mother's tolerance for intrusions upon the homesite. During the lactation period, usually gregarious female gray squirrels, *Sciurus carolinensis*, often defended the den trees in which their offspring were located (Sharp, 1959). In some species, the location of the first parturition site may have lasting effects. Lockly (1961) observed that female European rabbits tended to litter in the same site season after season despite the availability of superior locations. It seemed as if experience with the first litter established a "caretaking site."

The effects of the young on the parents' relationship to and utilization of their habitat may be quite direct; Kawai (1963) documented how a young Japanese macaque's responses to food which was provided by humans led to such changes. In an island-dwelling troop, an 18-month-old female monkey was the first to use water to wash sand off the sweet potatoes which the experimenters deposited on the beach. Within three months, the mother of this infant and three of her playmates displayed the same behavior. The habit led indirectly to an increase in the range of the group; as sweet potato washing spread, the monkeys began to eat univalves from a rock on the beach. Within six years, the young monkeys were not only playing on the beach and in the water, but were entering the sea in search of food — despite the fact that the members of the troop had previously shunned the ocean. In a later report, Kawai (1965) described the innovation and propagation of yet another use of water — to separate sand from grains of wheat. This behavior was introduced by the same female (then 4 years old) and spread among her playmates and from them to their older sibs and mothers.

3. Feeding Behavior

The foregoing examples suggest that just as the feeding behavior and food preferences of the offspring may be influenced by their parents, the young can affect the dietary habits and foraging techniques of their caretakers. Support for this contention was provided by Itani (1958), who followed the course of another troop's coming to accept and eat candies offered by humans. Here, too, it was found that adults tended to acquire the response in "imitation" of the infants with whom they came in contact, either as the primary caretakers, or while they were the loci about which juvenile play groups formed.

The intensity as well as the extensity of maternal feeding is affected by caretaking. In the early postpartum weeks there were marked increases in the food intake of primiparous laboratory rats (Grosvenor, 1956; Ota and Yokoyama, 1967a) and bank voles, *Clethrionomys glareolus* (Kacz-marski, 1966). In both Ota and Yokoyama's and Kaczmarski's studies, the mothers' rates of food consumption were found to vary with changes in litter weight-gain. When litter size was varied from 2 to 4, 8 or 12 pups, maternal food consumption increased according to offspring nutritive requirements (Ota and Yokoyama, 1967b). When litters were removed at day 12 and foster litters presented on day 17 postpartum, there was a brief decline in maternal food intake which was followed by an immediate enhancement of appetite when litters were returned (Ota and Yokoyama, 1967a).

Not only may the amount of forage consumed be affected, but in predatory species, the customary mode of hunting and/or pattern of consummation may differ between caretakers and noncaretakers. On two occasions, a mother tiger, *Panthera tigris*, failed to kill a tethered buffalo, apparently withdrawing as soon as the prey was down in order to allow her yearling cubs to make the kill (Schaller, 1967). In grass-hopper mice, *Onychomys leucogaster*, the parental eating pattern was observed to change when the offspring were between 10 and 16 days old. After the young reached this age, the abdominal parts of captured in-sects, which usually were devoured by adults, were frequently left in the nest by the parents (Ruffer, 1965).

4. *Dispersal*

Although the most commonly observed pattern is for adults to drive the young away as they near maturity (e.g., Wynne-Edwards, 1962), the young of some species may facilitate the relocation or social retintegra-tion of the adults (caretakers). Both Lockley (1961) and Southern (1940) observed that pregnant wild rabbits dug fresh stops away from the main tunnel system. Later, as they matured, the young enlarged these bur-rows, often connecting them with the main warren. On the other hand, in some species, the young have been observed to remain at their birth-sites while the adults depart. Davis *et al.* (1962) suggested that the matur-ation of young Mexican free-tailed bats, *Tadarida condylura*, caused over-crowding in the nursery caves and thereby led to adult dispersal. Each of four mother tree squirrels, *Tamiasciurus*, either surrendered a corner of their original territories to their weanlings or established a new territory altogether (Smith, 1968). King (1955) reported that the extreme gre-gariousness of newly emerged black-tailed prairie dog pups, *Cynomys ludovicianus*, caused the adults to leave the existing burrow system and construct a new one nearby.

In summary, the data indicate that mammalian offspring may directly affect their parents' movements in space, choice and utilization of home-sites, and feeding behavior and cause them to colonize uninhibited areas.

B. RELATIONSHIP OF THE PARENT TO THE ANIMATE ENVIRONMENT

1. Predators

Among those species in which the caretaker either remains with the offspring or actively defends the young against potential predators, the likelihood of parental exposure to this form of danger is increased during the period of immature dependence. For instance, an entire wolf pack defended its den and pups against a bear which, in a food-competition situation, would have been avoided (Murie, 1944). Mothers of several cetacean species have been reported to remain with and/or defend their calves despite imminent danger (Caldwell and Caldwell, 1966). The sight of an infant vervet or brazza monkey in human hands provoked more threats and much closer approach to man from adult conspecifics than would other circumstances (Booth, 1962). The tendency of mother Thompson's gazelles, *Gazella Thompsonii,* to return to or remain in the vicinity of their concealed fawns made them more vulnerable to attack by African wild dogs (Estes and Goddard, 1967). Caretaking may also indirectly increase the exposure of the parent. Schaller (1967) reported that the food demands of a litter of growing tiger cubs forced their mother to widen her hunting range to include areas near human habitation.

2. Conspecifics

a. *Gregariousness.* As was indicated by the above-mentioned role of rabbit young in extending the maternal nest-stop, the advent of young may affect the intraspecific social relationships of the caretaker. According to Eisenberg (1966, p. 18), the parent-offspring relationship is a "fundamental social unit" in mammals which constitutes one of the major means whereby integrated social organizations are formed. In so-called solitary forms, parental care may represent the only long-term social interaction in which adults participate (see Eisenberg, 1966, for a review). Among seasonally gregarious species such as elk and moose (Altmann, 1963) or in those which form only transitory group associations, e.g., chimpanzees (Goodall, 1965), the mother-offspring relationship may constitute the only regular, enduring conspecific interaction.

In a number of group-living primate species, it has been suggested that the presence of young may be one of the most potent cohesive forces in group maintenance (De Vore, 1963; Jay, 1963; Jolly, 1966a), and that the young adult male's attraction to offspring may be one determinant of the form of the social group (Kummer, 1967). Thus, the caretaker's long-term gregariousness as an adult may be in part a response to the presence of young.

However, among typically social species, the immediate and/or impending presence of offspring may be related to changes in group structure. In some mammals, females withdraw from the group in order to give birth. The extent of this separation varies across species. For example, the separation of American bison, *B. b. bison* (McHugh, 1958), or sifak lemurs, *Propithecus verreauxi* (Jolly, 1966a), is relatively brief, lasting at most 2 days. In contrast, female blackbuck, *Antilope cervicapra*, remain apart from the herd from parturition until 2 weeks postpartum (Schaller, 1967), and coatis, *Nasua narica*, leave the band in order to give birth and remain apart for up to 40 days postpartum (Kaufmann, 1962).

b. Peer Relationships. Where mothers remain with or return to their groups, the presence of the young may be related to modifications of the pattern of interrelationships therein. While the adult females within male-dominated groups of hamadryas baboons were usually kept apart from members of other units by their leaders, brief, tentative relationships arose between adult and juvenile females from nearby units and mothers of small infants. These brief associations, and the formation of unusually large and fluid groups of mothers with small infants, were considered to be a consequence of the females' attraction to each other's infants (Kummer, 1968). In a captive colony of rhesus monkeys, nulliparous females were often observed to intensify reciprocal grooming relationships with new mothers, apparently in order to gain access to the latters' offspring (Hinde and Spencer-Booth, 1967a; Rowell *et al.*, 1964). Similar increases in the amount of contact between females were re-reported in captive dolphins, *Tursiops Truncatus* (Tavolga and Essapian, 1957), and in captive colonies of bushbabies, *Galago senegalensis bradfieldi* (Sauer, 1967), after the birth of young.

Such "aunt"-infant interactions may lead to the formation of new interadult bonds. In a colony of captive squirrel monkeys, *Saimiri sciureus*, Ploog (1967) observed that a relationship was occasionally formed de novo between two females several weeks after one of them had given birth, apparently due to the other's interest in the infant. Further, in this colony, such relations were observed to persist beyond the phase of caretaking (see also Rosenblum, 1968). In a laboratory colony of Patas monkeys, *Erethrocebus patas*, the birth of an infant led to a marked change in the behavior of a nonlactating female toward a new mother. Aggressive approaches were replaced by "friendly" overtures, and, while the mother had previously initiated most of the grooming bouts, the other female came to be the initiator. Further, the mother's threats were no longer responded to with aggression (Hall and Mayer, 1967).

The development of new intragroup relationships is not always limited to interactions between females. In some troops of Japanese macaques, subdominant males "adopted" yearlings which were displaced by the birth of a sibling. In several cases in which a female infant was adopted, "paternal" care led to enduring grooming relationships between the males and the females for whom they had provided care (Itani, 1959). In hamadryas baboons, Kummer (1968) has suggested that the characteristic one-male groups resulted from a young adult male's adopting a caretaking role toward a juvenile female; as the female follower matured, caretaking elements gradually became replaced by sexual and leadership behavior.

The role of the caretaker in the wider group may be affected by the presence of or its solicitude for the young. These effects include food-gathering behavior; a mother wolf remained at the den with her pups while the litter was young and did not accompany the pack on the hunt (Murie, 1944). Intergroup conflict also may be modified; in the wild, mother langurs, *Presbytis entellus,* did not participate in the troop's aggressive displays or attacks against other troops (Ripley, 1967). Likewise, patterns of intragroup contact may change with the advent of young; the male in a laboratory colony of squirrel monkeys avoided a female during labor and lactation (Bowden *et al.,* 1967). The cohesion of subgroups within the wider group may be affected; female hamadryas baboons which had newborn infants, compared to the other females, were permitted to stray farther from their group leaders (Kummer, 1968). Among free living savannah-dwelling baboons, *Papio cynocephalus,* females with infants received more protection from dominant males in the troop than did nullipara (De Vore, 1963).

c. Social Rank. Several of the foregoing examples suggest that, in gregarious species, caretaking may affect certain aspects of an individual's social rank. This seems to be a fairly widespread occurrence at least with respect to rights to the young, and sometimes to other commodities. In captive barbary apes, *Macaca sylvana* (Lahiri and Southwick, 1966), feral langurs (Jay, 1963), free-ranging sifak lemurs (Jolly, 1966a), and feral mountain gorillas, *Gorilla gorilla beringei* (Schaller, 1963), females with young infants had higher dominance status than those with older or no offspring. Similar observations have been reported for subprimate forms. Barren ground caribou cows with calves were more dominant than cows with yearlings or no offspring (Lent, 1966), and mother raccoons, *Procyon lotor*, with the young of the year were dominant over other age-sex classes at a winter feeding station (Sharp and Sharp, 1956).

In one instance, a change in dominance has been related to a physiological concomitant of parenthood. While male and nulliparous female reindeer, *Rangifer tarandus* L., lost their antlers, pregnant and lactating females did not, and thus gained relatively higher status (Espmark, 1964).

Where caretaking extends beyond the parent-offspring relationship, other individuals may also be affected. According to Itani (1959), subdominant male Japanese macaques which adopted displaced yearlings were granted access to the central area of the troop—an area which was normally the exclusive prerogative of the most dominant males.

In some primates, the offspring may play an active role in modifying caretaker rank. By manipulating group composition, Marsden (1968) studied the agonistic interactions of a group of 10 rhesus monkeys in a large enclosure. Not only did the rank of the infants vary with that of their mothers, but, during several periods of reorganization, the presence or absence of support by the young accounted for at least one shift in the mothers' standings in the dominance hierarchy.

d. Aggression. Changes in social dominance suggest that there may be alterations in aggressive relations. Several reports have related the performance of aggressive behavior to the presence of offspring. Hall and Mayer (1967) observed that 45% of all the instances of threats between two mother patas monkeys occurred in response to excessively "rough" play of the other's infant. Further, two-thirds of all the threats that these females directed toward the dominant male were in response to incidents involving their offspring. Male lions', *Felis leo*, occasional intolerance of cubs at the kill often aroused defensive threats from their mothers (Guggisberg, 1961).

e. Communication. To the extent that the young may trigger particular signaling behavior in caretakers (e.g., Lent, 1966; Nelson, 1965), their advent or presence affects adult communicative behavior. In some gregarious species, the communicative behavior of the entire group may be altered. Adult domestic dogs in a kennel tended to reduce their vocalizations whenever a bitch was whelping (Bleicher, 1962), and Joslin (1967) reported fewer howls from a wild wolf pack while a litter of pups was young.

f. Reproduction.   Reproductive activities themselves can be affected by caretaking. It is a common observation that sexual receptivity in females of a number of species is delayed until the cessation of nursing, or at least beyond that period during which nonnursing females resume cycling (e.g., cf. Meites, 1966). During the period in which lactation can be extended by repeated litter-replacement in rats (Bruce, 1961; Nicoll and Meites, 1959) and mice (Selye and McKeown, 1934), the anestrous condition is sometimes comparably prolonged. Recent evidence suggests that, in rats, lactation diestrus can be affected by non-suckling stimuli afforded by the litter (Moltz *et al.*, 1969b; see also Rothchild, 1967).

Although precise experimental data are lacking, it seems that the phenomenon of lactation diestrus may be fairly widespread among mammals. There are reports of this condition in domestic sows, *Sus domesticus* (Crighton, 1967), short-tailed shrews, *Blarina brevicauda* Say (Pearson, 1944), longtailed weasels, *Mustela frenata* (Wright, 1948), Weddell seals, *Leptomychotes weddelli* (Ray, 1967), rhesus monkeys (Vandenbergh and Vessey, 1966), and chimpanzees (van Lawick-Goodall, 1967). In these forms, then, the presence of the young may be said to influence the display of their caretaker's mating behavior. Direct effects of the presence of offspring on mating are not limited to the mother's endocrine status. According to Altmann (1958), during the rut, the bull moose *Alces alces*, must respond at least in a "neutral" manner to a cow's calf in order to be accepted by her.

Caretaking itself may be affected by exposure to the young above and beyond the obvious fact of its elicitation; prior caretaking often modifies responses to subsequent offspring. While multiparous, ovariectomized, cesarean-delivered Sprague-Dawley rats accepted and reared foster litters, 18 hours after operation, only half as many identically operated primipara did so (Moltz and Wiener, 1966). Similarly, when compared with a group of comparably treated primipara, twice as many multiparous, progesterone treated, cesarean-delivered females accepted foster litters 24 hours after surgery (Moltz *et al.*, 1969a). A similar effect has been observed in rhesus monkeys. As noted above, Seay *et al.* (1964) and Arling and Harlow (1967) observed that isolation-reared female rhesus monkeys usually failed to display caretaking behavior toward their first-born infants. However, six of seven such females manifested a high degree of solicitude for their second offspring insofar as they allowed their infants to make more ventroventral contact and engaged in more play with them than did experienced feral-reared mothers (Harlow *et al.*, 1966). Prior caretaking experience has also been found to reduce as well as facilitate the display of certain aspects of parental response in rodents (Noirot, 1964; Richards, 1967) and primates (Seay, 1966). Thus, the reproductive behavior of caretakers, like their affiliative, agonistic, and communicative activities, may be altered as a result of prior or current caretaking experience.

These data show that the presence or prior bearing of young affect modifications in the behavior of the caretaker, and that such changes can be demonstrated in every realm in which traditional formulations have emphasized the "educative" effect of parents on the young.

IV. THE ROLE OF THE OFFSPRING IN THE ADAPTIVE RADIATION OF THE SPECIES

The evidence reviewed in the first section indicates that the necessity for caretaker tutelage in the behavioral development of the young may have been overestimated, and the data presented in the preceding section show that the effects which are exerted are often symmetrical. These findings suggest a reformulation of the evolutionary significance of the mammalian parent-offspring relationship.

The most commonly accepted phylogenetic function of the association of parent and offspring, beyond nursing, is to allow the immature individual to acquire a number of behavior patterns through "education" while its nervous system is still developing and before it must fend for itself (Romer, 1958, 1967). Education has been seen as conferring an advantage in that locally adaptive traditions may be transferred from one generation to the next (e.g., Hamburg, 1969; Washburn and Hamburg, 1965; Wynne-Edwards, 1962). The other functional role proposed for the phenomenon of postnatal caretaking is that parental care allowed for the evolution of direct population-regulating mechanisms such as litter desertion and cannibalism (Wynne-Edwards, 1962). Both these formulations are concerned with species adaptations within a particular habitat. In this section I shall suggest that the parent-offspring relationship, in at least some mammalian forms, may also function to increase the likelihood that new ecological opportunities will be utilized.

A. PARENTAL CARE AND THE RANGE OF POTENTIAL ADAPTATIONS

1. The Effects of Viviparity and Postnatal Care on Offspring Response-Variability

Simpson (1950) has proposed that one of the most important features in the continuing evolution of a class is the range and variety of adjustments that individuals can make to their environments. According to

Dobzhansky (1951), those forms in which parental care is most highly developed appear to be the most rapidly evolving groups in their respective phyla. Taken together, these views imply that caretaking may lead to an enhancement of flexibility of response to the environment. If the evolution of embryonic development within the uterus, followed by postpartum care and milk feeding, served to minimize environmentally induced physical variation (Simpson, 1950), adaptivity could have been further enhanced by an increase in the potential variability of the offspring's behavioral responses to fluctuations in their surroundings without adversely affecting fitness. For instance, in various inbred strains of laboratory mice, Ginsburg and Jumonville (1967) have shown that the effects of "stressors" administered in the early postnatal period vary according to the genotype of the subject. Since each strain represents a genotypic possibility available in the parent population, such interstrain variability would reflect part of the range of adaptive potential within that heterogeneous population. A wide range of individual reactions would enhance species survival by increasing the probability that some individuals' responses would be adaptive. The data reviewed in this paper suggest that there may be at least two behavioral areas in which the adaptive potential of mammals may have been increased by this means. The first would include the choice of habitat—e.g., Wecker's (1963) finding in prairie deer mice that the propensity to acquire habitat preferences seemed stronger than the inherent tendency to seek open areas. The second would include the fixity of response to the details of many of the animate and inanimate objects in its environment—e.g., the tendencies to explore and play (pp. 77, 83 and below).

2. Exploration and Play as Favoring Innovation by the Young

a. *Exploration.* A number of studies suggest that young mammals are more attracted to their surroundings than are adults. In a study of the responses of Japanese macaques to inanimate objects, Menzel (1966) found that 81 of 111 recorded instances of an animal's manipulating an object involved infants in their first year; in 24 additional cases, the animals were in the neighborhood of 2 years of age. Only 6 of the 111 instances of object manipulation involved animals over the age of three years. Since the infants in the troops under observation amounted to only 30–40% of the population, but accounted for over 70% of the object-contacts, it seems clear that the young showed a greater propensity to actively investigate the stimuli. Further, when caches of test objects were "robbed" by the monkeys, infants were the ones who were seen in possession of the pilfered goods on the following morning.

Similar findings have been reported for other species. When the motor behaviors of 230 feral rhesus monkeys on Cayo Santiago island were recorded over a 2-month period, the younger animals more frequently manipulated objects and spent more time in the trees (Draper, 1966). While adult feral hamadryas baboons seldom manipulated nonedible objects, the 1-year-olds showed the highest frequency of this behavior (Kummer, 1968). Welker (1956) tested the responsiveness of three 3–4 and three 7–8-year-old captive chimpanzees to trays containing a variety of objects. The younger animals were generally more responsive in that they investigated the objects for longer periods. Such curiosity is not limited to young primates. The investigatory behavior of domestic cattle, *Bos taurus,* has been reported to be inversely related to age (Hafez and Schein, 1962), and in each of four hand-reared young raccoons, exploratory and playful responses to inanimate objects predominated during the animals' "childhoods" (Welker, 1959). Martinek and Lat (1968) found that, in an open field, young dogs showed more general activity, including rearing, moving about, and sniffing, than did adult dogs. Puppies also alternated their type of activity more often, and their activity habituated more slowly than that of adults.

In addition to being attracted to their surround, the young may be less likely than adults to form avoidance habits. Campbell (1967) exposed rats to a fear-conditioning situation at 18, 23, 38, 54, or 100 days of age. Animals from each training group were tested immediately, 7, 21, or 42 days after the conditioning trials. The two younger age groups displayed markedly less retention than the older animals. Campbell suggested that this phenomenon might be adaptive in that only consistently dangerous situations would lead to enduring avoidance tendencies and the early development of stereotyped responses to essentially fortuitous environmental coincidences would be avoided. Thus, the younger animals' potential for approaching novelty is more likely to be given full expression.

Data on geographical mobility suggest the adaptive potential of the offspring's greater tendency to investigate. Howard (1949) found that female deermice tended to frequent the same general area only after sexual maturity, and Lowe (1966) reported that immature red deer, *Cervus elaphus* L., on the Scottish island of Rhum, were much more geographically mobile than adults.

Greater mobility of juveniles in the social environment is also a common observation. Sugiyama (1960) recorded the events surrounding the division of a free-ranging troop of Japanese monkeys. In this process the juvenile's greater adaptibility to changes in group structure and home-site was found to play a major role. Similarly, Kummer (1968) reported that while adult hamadryas baboons (whose social organization centers about a single male and his jealously guarded following of one or more females) devoted less than 10% of their social interactions to members of other groups, infants and juveniles spent twice as much time in extra-group contact. This tendency, too, is not limited to primates. Koford (1957) observed that among highly territorial bands of vicuna, juveniles would leave their own groups for play.

b. The Adaptive Function of Play. The behavioral plasticity of mam-mals is particularly evident in the play of the young. In play, stereotyped motor patterns are elicited by a wider range of objects and often seem to be "detached" from responses accompanying the same acts in adult sequences (Eibl-Eibesfeldt, 1967; see also Jewell and Loizos, 1966; Thorpe, 1966). Since there is ample evidence that relatively novel ob-jects increase the tendency of young mammals to engage in such be-havior (e.g., Mason, 1965), the comparative autonomy of otherwise stereotyped motor acts and the tendency to direct them to a wide variety of animate and inanimate objects would seem to be at least potentially valuable for adapting a species to changing or unexploited surroundings.

c. The Offspring as Innovators. In Section II, a number of findings were cited which indicated that the young played an active role in making contact with, and introducing their parents to, aspects of the environ-ment. Kawai's (1965) analysis of genealogical records showed that the sweet potato washing habit in Japanese macaques spread from the younger members of the troop to older individuals through contact in play groups or caretaking interactions. Kawai suggested that transmis-sion in the "expected direction" from caretaker to young occurred only after the response was acquired by adults or when the innovators matured and were caring for their own offspring.

These data indicate that it is the preadult age class which plays a major role in innovative activity. Indeed, the Japanese investigators have carefully documented observational and experimental evidence in support of this view. Kawai (1965) reported that in the first five years after the introduction of sweet potato washing, only two of eleven adults acquired the response, while 15 of 19 animals which were two years old or younger at the time of innovation had done so. Like feeding patterns associated with water, the tendency to enter the sea, either in play or to retrieve peanuts deliberately thrown into the water, was found to be most readily acquired by juveniles. While about 96% of the animals born within three years of the introduction of "bathing" entered the water, only 26% of those which were already adults did so.

In an earlier experiment with another free-ranging troop of Japanese macaques, Itani (1958) tested the responses of various age groups to the presentation of candy. After the end of six presentations of candy, while only 32% of the adult males and 31% of the adult females had accepted his offerings at any time, over 90% of the monkeys which were under the age of four years accepted and ate the sweets.

Tsumori (1967) has extended the analysis of age differences in the acquisition of novel behavior to include an evaluation of task complexity. As a test of the animals' "adaptability," he buried peanuts under sand while being watched by the monkeys. In this situation, which required sustained attention and some problem-solving skill, it was found that the older juveniles and early adolescents were most successful. As in the previous studies, however, the adults were less likely to acquire the novel sand-digging response, although all age-classes were attracted to the bait.

Suzuki (1965) studied the adaptations of Japanese macaques to different climatic conditions. During the course of his investigations of the behavior of a small troop which lived in a very cold region he observed, for the first time, a juvenile female "bathing" in a hot spring. Within two months, five of the eight troop members which were between the ages of one and four years entered the spring. In the following winter six infants also acquired the habit; no adults were observed bathing.

From the foregoing, it seems likely that one function of the greater responsiveness of the young to their environment would be to introduce new habits or to exploit new habitats. Of course, adults can and do introduce new patterns on occasion. Suzuki (1965) also reported that the oldest female in the same hot-spring-bathing troop, mentioned above, introduced a pattern of washing apples which were thrown to the monkeys by the investigators. However, this is the only recorded instance known to the writer in which an adult was the innovator, as compared to several recorded innovations introduced by juveniles. Since the juvenile class usually comprises less than one half of the average troop, the apparently greater frequency of innovation by the young would seem to reflect a genuine age difference. Although the only well documented instances of such propagation known to the writer are those summarized above in Japanese macaques, it seems reasonable to suppose that similar phenomena occur in other species—particularly in gregarious animals in which exploration and play characterize the activity of the young.

B. Parental Care and Adaptive Radiation

The bulk of the evidence surveyed above indicates that adults (caretakers) are often the more "conservative" elements in mammalian populations and that usually they transmit only established group tradition (albeit largely unwittingly). Although the propensity to emulate the behavior patterns performed by their caretakers is a characteristic of many mammalian young, such behavior adapts individuals primarily to their existing habitats. While advantageous for individual survival, such modeling functions as a conservative mechanism. If ". . . expansion into new adaptive zones is the rule of increase for any group" (Simpson, 1950, p. 114), the continuing evolution of mammalian species seems more likely to have derived from the offspring's introducing and propagating novel responses.

Innovation involves risk. However, most species which have adapted to a given habitat can produce more young than are necessary to maintain an optimum population density (Wynne-Edwards, 1962). Thus, such a species can afford to beget offspring whose behavior is relatively unstructured (e.g., play) and who are more attracted to the details of the environment (e.g., exploration) so long as it stands to gain from the introduction of new means of exploiting its surroundings. Whether successful or fatal, innovations attempted during the period of close adult-offspring association will supply "information" to the adults and to the young individual's peers. The caretaker's contribution is to provide the milieu in which chances will be taken. Successful innovations will be transmitted "upward" to the caretakers who associate closely with the offspring, "horizontally" to playmates, and "downward," becoming tradition, when and if the innovator assumes a caretaking role.

Data consistent with this suggestion are provided by an experiment by Menzel (1966). When he placed a doll atop a food-kettle from which Japanese monkeys had been accustomed to feed, he found that the troop members' approach responses were inhibited. With the doll present, the adults' feeding seemed to be dependent upon the juveniles' initiative, either directly, when the young were the first to approach, or indirectly, by the adults' stealing the food obtained by the juveniles. From his report, the adults seemed to let the juveniles "take the risks."

Thus, it is argued that, in addition to the genetic "novelty" and variability inherent in sexual reproduction (cf. Dobzhansky, 1951), mammals evolved additional behavior patterns, peculiar to the period of immaturity, which enhanced the likelihood of adaptive radiation. Available evidence suggests that, in lieu of stimulus-specific approach and manipulatory-ingestive responses, the trend of mammalian evolution may have been toward the offspring's displaying generalized approach tendencies and expressing novel combinations of species-typical motor patterns.

These trends were probably accompanied by the establishment of a mechanism whereby prepotent (cf. Hediger, 1955), generalized (cf. Thorpe, 1962), flight reactions could be blocked or reduced in the offspring by the presence of a familiar caretaker and thus allow the full expression of investigative and play responses in situations which were otherwise neutral. Liddell (1954, 1960) demonstrated the importance of the mother's proximity in modulating flight and emotional disturbance in sheep and goats. When trained in the absence of their dams, lambs and kids subjected to a shock-conditioned leg withdrawal became agitated and hyperreactive—even when away from the laboratory. In contrast, animals whose mothers were present during similar training displayed much less disturbance and reacted more freely. Apparently, early exposure to such training while with the mother also reduced the intensity of the long-term sequelae of early conditioning; as adults, goats which were trained during infancy in the presence of their dams were less subject to "experimental neurosis" than were animals which had been similarly treated while separated from their mothers (Liddell, 1954, 1960). The facilitating effects of the presence of an inanimate mother-surrogate on the investigative and play responses of infant rhesus monkeys (Harlow, 1961, 1962) has already been mentioned.

Thus, among several contemporary mammalian species, the caretaker appears to provide the milieu which assures not only the expression of physiological maturation and species-typical social behavior, but facilitates the display of novel and idiosyncratic modes of action in the environment. The offspring contribute the variability of response which provides the basis for further adaptive radiation of the species, and the relationship between caretaker and young affords a mechanism whereby earlier generations may profit from successful innovations.

The parental role in enhancing the responsiveness of the offspring appears to be more permissive than active. For example, although cloth-surrogate mothers ". . . are devoid of facial, vocal and gestural language; cannot protect their infants; cannot punish their infants; and cannot respond reciprocally to their infants' behavior . . ." (Harlow and Harlow, 1969, p. 28), rhesus monkeys raised with them will venture forth to explore, play with, and even destroy objects which otherwise would have left them immobilized by fear (cf. Harlow, 1961).

In summary, the data reviewed in Section I of this paper indicate that mammalian young frequently develop species-typical behavior patterns in the absence of opportunities to interact with conspecific caretakers. The studies summarized in this section show that the mother (or mother surrogate) often has a profound effect on her offspring's behavior simply by virtue of her presence. Thus, in addition to fostering an under-estimation of the evolutionary significance of offspring behavior, the focus on parental "shaping" or "tuition" may emphasize relatively un-important processes at the expense of more significant, although largely passive, facilitating effects. Conventional views of the functions of the parent-young relationship in mammals may therefore attribute too much significance to essentially conservative processes and pay too little attention to the analysis of the more or less passive ways in which the parent provides an optimal setting for offspring behavioral development. It seems that, ontogenetically and phylogenetically, the behavioral role of the parent vis-a-vis its young is as permissive as it is constitutive.

C. IMPLICATIONS

These evolutionary speculations can give rise to testable hypotheses. For instance, it would be expected that, within a related group, the young of the more widely distributed subspecies will display more exploratory-investigatory behavior and/or idiosyncratic modes of deal-ing with their surroundings than the offspring of their less widely dis-tributed relatives. Further, where members of a species develop local traditions, not only should the species be less specialized, but the young would be expected to show more novel response patterns than the off-spring of related forms whose adult behavior is more stereotyped. Simi-larly, the argument presented here leads to the prediction that the (environmentally oriented) behavioral development of the young of more specialized forms would be less subject to environmentally in-duced variation than that of related, less specialized species.

The experimental and observational data reviewed in this paper make it clear that, behaviorally speaking, the functional significance of the parent-offspring relationship is considerably broader and more complex than that suggested by the model of parental tuition. In particular, social facilitation—e.g., the mother cat exciting her offspring to deliver the killing bite (Leyhausen, 1968)—has received less attention than it de-serves. Perhaps more important, especially in species which defend their young, is the need to understand the mechanisms by which the mere presence of a caretaker reduces the offspring's flight tendencies (Har-low, 1961, 1962; Liddell, 1954, 1960). Finally, the data indicate an im-perative need to appreciate the active contribution of the young to their own development (Ewer, 1969; Mason, 1968) and to determine the nature of the often very general environmental conditions (e.g., Harper, 1968a; Mason, 1968; Thoman and Arnold, 1968b) necessary for the development and display of normal behavior.

Summary

1. Traditional views of the functions of the parent-offspring relationship in mammalian ontogeny and phylogeny are reviewed and subjected to a critical evaluation.

2. A substantial body of evidence indicates that, during the period of maximal parent-offspring interaction, modifications occur in the responses of the young toward their physical and social environments.

3. However, other data suggest that behavioral development often results from somewhat different processes than those suggested by such terms as "education" or "shaping." It seems that, in many forms, prior experience with caretakers is not necessary for the development of the offspring's tendencies to display gregariousness, species-typical social responses or exploratory-manipulative behavior.

4. The activity of the offspring in their physical and social surroundings appears to be facilitated by the mere presence of the caretaker, which (particularly in those species that actively defend their young) apparently suffices to inhibit or reduce the offspring's tendency to flee from novel objects.

5. Phylogenetically, the adaptive advantage customarily attributed to the parent-offspring relationship is the transmission of locally adaptive "traditions" from one generation to the next. This view is most often associated with the assumption that the young depend upon social interaction for many species-specific behavior patterns. The data reviewed in this paper provide only limited support for such a contention.

6. In view of the fact that the transmission of tradition is essentially conservative, whereas mammalian evolution seems to be largely the result of the exploitation of new habitats, a new look at the problem is required.

7. Evidence is available which indicates that, for each general class of offspring response which is affected by the association of parent and young, corresponding modifications in caretaker behavior can be demonstrated (often in the same species).

8. Evidence from a number of species indicates that the young are the segment of the population which is the most likely to introduce — or "experiment with" — novel behavior patterns. Thus, the innovative

activity of the juvenile class may have played a major role in the adaptive radiation of at least some mammalian species.

9. The caretaker-offspring nutritive and defensive relationships may have provided a mechanism whereby the threat of environmental disturbances of species-typical development was reduced. These exogenous safeguards of development could have allowed the maintenance of an acceptable survival rate despite an increase in the behavioral flexibility of the young individual's responses (exploration and play) to its surroundings.

10. Thus, the response-variability of the young could be increased while generalized flight tendencies became subject to inhibition by the presence of a familiar caretaker. Concomitantly, maturation of the tendencies to maintain habitual patterns and to be wary of the novel was delayed until adulthood.

11. Combined with a reproductive capacity which permitted recruitment in excess of the level required to maintain the species, these trends may have resulted in the mammalian young acting as "pioneers" in the adaptive radiation of this Class.

12. It is concluded that emphasis on the "educative" function of the parent has led to an underestimation of the role of offspring behavior in mammalian evolution and an exaggeration of the significance of parental tuition in mammalian behavioral development.

Acknowledgment

I am indebted to Richard Q. Bell and Paul D. MacLean for advice and encouragement and to Gale Inoff for assistance in the preparation of this paper.

References

Alexander, B. K., and Dodsworth, R. O. 1966. Peer deprivation in mother-reared rhesus monkeys. Mimeo., University of Wisconsin, Madison, Wisconsin.

Alexander, B. K., and Harlow, H. F. 1965. Social behavior of juvenile rhesus monkeys subjected to different rearing conditions during the first six months of life. *Zool. Jahrb. Physiol.* **71**, 489–508.

Altmann, M. 1958. Social integration of the moose calf. *Anim. Behav.* **6**, 155–159.

Altmann, M. 1963. Naturalistic studies of maternal care in moose and elk. *In* "Maternal Behavior in Mammals" (H. L. Rheingold, ed.), pp. 233–253. Wiley, New York.

Arling, G. L. 1966. Unpublished M.A. Thesis, University of Wisconsin, Madison, Wisconsin.

Arling, G. L., and Harlow, H. F. 1967. Effects of social deprivation on maternal behavior of rhesus monkeys. *J. Comp. Physiol. Psychol.* **64**, 371–377.

Avery, G. T. 1928. Responses of foetal guinea pigs prematurely delivered. *Genet. Psychol. Monogr.* **3**, 249–331.

Bartholomew, G. A., and Hoel, P. G. 1953. Reproductive behavior of the Alaska fur seal, *Callorhinus ursinus. J. Mammal.* **34**, 417–436.

Beach, F. A. 1942. Comparison of copulatory behavior in male rats raised in isolation, cohabitation and segregation. *J. Genet. Psychol.* **60**, 121–136.

Beach, F. A. 1958. Normal sexual behavior in male rats isolated at fourteen days of age. *J. Comp. Physiol. Psychol.* **51**, 37–38.

Beach, F. A. 1968. Coital behavior in dogs. III. Effects of early isolation on mating in males. *Behaviour* **30**, 218–238.

Berkson, G. 1968. Development of abnormal stereotyped behaviors. *Developm. Psychobiol.* **1**, 118–132.

Blauvelt, H. 1956. Neonate-mother relationships in goat and man. *Group Processes, Trans. 2nd Conf., 1955* pp. 94–140.

Bleicher, N. 1962. Behavior of the bitch during parturition. *J. Amer. Vet. Med. Ass.* **140**, 1076–1082.

Booth, C. 1962. Some observations on behavior of Cercopithecus monkeys. *Ann. N.Y. Acad. Sci.* **102**, 477–487.

Bowden, D., Winter, P., and Ploog, D. 1967. Pregnancy and delivery behavior in the squirrel monkey (*Saimiri Sciureus*) and other primates *Folia Primatol.* **5**, 1–42.

Bronson, G. W. 1968. The development of fear in man and other animals. *Child Develop.* **39**, 409–428.

Bruce, H. M. 1961. Observations on the suckling stimulus and lactation in the rat. *J. Reprod. Fert.* **2**, 17–34.

Butler, R. A. 1965. Investigative behavior. *In* "Behavior of Nonhuman Primates" (A. M. Schrier, H. F. Harlow, and F. Stollnitz, eds.), Vol. II, pp. 463–493. Academic Press, New York.

Cairns, R. B. 1966. Attachment behavior of mammals. *Psychol. Rev.* **75**, 409–426.

Caldwell, M. C., and Caldwell, D. K. 1966. Epimeletic (care-giving) behavior in cetacea. *In* "Whales Dolphins and Porpoises" (K. S. Norris, ed.). Univ. of California Press, Berkeley, California.

Calhoun, J. B. 1962. The Ecology and Sociology of the Norway Rat. U.S. Public Health Service Publ. No. 1008. Bethesda, Maryland.

Campbell, B. A. 1967. Developmental studies of learning and motivation in infra-primate mammals. *In* "Early Behavior" (H. W. Stevenson, E. H. Hess, and H. L. Rheingold, eds.), pp. 43–71. Wiley, New York.

Carpenter, C. R. 1940. A field study in Siam of the behavior and social relations of the gibbon (*Hylobates* lar.). *Comp. Psychol. Monogr.* **16**, 1–212.

Cockrum, E. L. 1956. Homing, movements and longevity of bats. *J. Mammal.* **37**, 48–57.

Collias, N. E. 1956. The analysis of socialization in sheep and goats. *Ecology* **37**, 228–239.

Collias, N. E. 1962. Social development in birds and mammals. *In* "Roots of Behavior" (E. L. Bliss, ed.), pp. 264–273. Harper & Row (Hoeber), New York.

Crighton, D. B. 1967. Effects of lactation on the pituitary gonadotrophins of the sow. *In* "Reproduction in the Female Mammal" (G. E. Lamming and E. C. Amoroso, eds.), Plenum Press, New York.

Davenport, R. K., and Menzel, E. W. 1963. Stereotyped behavior of the infant chimpanzee. *Arch. Gen. Psychiat.* **8**, 99–104.

Davenport, R. K., Menzel, E. W., and Rogers, C. M. 1966. Effects of severe isolation on "normal" juvenile chimpanzees. *Arch. Gen. Psychiat.* **14**, 134–138.

Davis, R. B., Herried, C. F., and Short, H. L. 1962. Mexican free-tailed bats in Texas. *Ecol. Monogr.* **32**, 311–346.

Denenberg, V. H. 1967. Stimulation in infancy, emotional reactivity and exploratory behavior. In "Neurophysiology and Emotion" (D. G. Glass, ed.), pp. 161–190. Rockefeller Univ. Press, New York.

Denenberg, V. H., Hudgens, G. A., and Zarrow, M. X. 1964. Mice reared with rats: modification of behavior by early experience with another species. Science 143, 380–381.

Denenberg, V. H., Hudgens, G. A., and Zarrow, M. X. 1966. Mice reared with rats: effects of mother on adult behavior patterns. Psychol. Rep. 18, 451–456.

De Vore, I. 1963. Mother-infant relations in free-ranging baboons. In "Maternal Behavior in Mammals" (H. L. Rheingold, ed.), pp. 305–335. Wiley, New York.

de Vos, A. 1960. Behavior of barren ground caribou on their calving grounds. J. Wildl. Manag. 24, 250–258.

de Vos, A., Brokx, P., and Geist, V. 1967. A review of social behavior of the North American cervids during the reproductive period. Amer. Midl. Natur. 77, 390–417.

Dobzhansky, T. 1951. "Genetics and the Origin of the Species." Columbia Univ. Press, New York.

Donaldson, S. L., Black, W. C., and Albright, J. L. 1966. The effects of early feeding and rearing experience on dominance aggressive and submissive behavior in young heifer calves. Amer. Zool. 6, 559–560.

Draper, W. A. 1966. Free-ranging rhesus monkeys: Age and sex differences in activity patterns. Science 151, 476–478.

Eibl-Eibesfeldt, I. 1961. The interactions of unlearned behaviour patterns and learning in mammals. In "Brain Mechanisms and Learning" (J. F. Delafresnaye, ed.), pp. 53–73. Thomas, Springfield, Illinois.

Eibl-Eibesfeldt, I. 1967. Concepts of ethology and their significance in the study of human behavior. In "Early Behavior: Comparative and Developmental Approaches" (H. W. Stevenson, E. H. Hess, and H. L. Rheingold, eds.). Wiley, New York.

Eisenberg, J. F. 1963. The behavior of heteromyid rodents. Univ. Calif. Berkeley, Pub. Zool. 69, 1–78.

Eisenberg, J. F. 1966. The social organizations of mammals. In "Handbuch der Zoologie" Vol. 8(10), pp. 1–92. Walter de Gruyter, Berlin.

Epple, G. 1968. Comparative studies on vocalization in marmoset monkeys (Hapalidae). Folia Primatol. 8, 1–40.

Espmark, Y. 1964. Studies in domanance-subordination relationships in a group of semidomestic reindeer (Rangifer tarandus L.) Anim. Behav. 12, 420–426.

Estes, R. D. 1967. The comparative behavior of Grant's and Thompson's gazelles. J. Mammal. 48, 189–209.

Estes, R. D., and Goddard, J. 1967. Prey selection and hunting behavior of the African wild dog. J. Wildl. Manage. 31, 52–70.

Evans, C. S. 1967. Methods of rearing and social interaction in Macaca nemestrina. Anim. Behav. 15, 263–266.

Ewer, R. F. 1963. The behaviour of the meerkat. Z. Tierpsychol. 20, 570–607.

Ewer, R. F. 1967. The behaviour of the African giant rat (Cricetomys gambianus Waterhouse). Z. Tierpsychol. 24, 6–79.

Ewer, R. F. 1969. The "instinct to teach." Nature 222, 698.

Folman, Y., and Drori, D. 1965. Normal and aberrant copulatory behaviour in male rats (R. norvegicus) reared in isolation. Anim. Behav. 13, 427–429.

Frädrich, H. 1965. Zur Biologie und Ethologie des Warzenschweines (Phacochoerus aethiopicus Pallas) unter Berucksichtigung des Verhaltens anderer Suiden. Z. Tierpsychol. 22, 328–393.

Fuller, J. L. 1967. Experiential deprivation and later behavior. Science 158, 1645–1652.

Fuller, J. L., and Clark, L. D. 1966a. Genetic and treatment factors modifying the post-isolation syndrome in dogs. *J. Comp. Physiol. Psychol.* **61**, 251–257.

Fuller, J. L., and Clark, L. D. 1966b. Effects of rearing with specific stimuli upon post-isolation behavior in dogs. *J. Comp. Physiol. Psychol.* **61**, 258–263.

Gerall, A. A. 1963. An exploratory study of the effect of social isolation variables on the sexual behavior of male guinea pigs. *Anim. Behav.* **11**, 274–282.

Gerall, H. D. 1965. Effect of social isolation and physical confinement on motor and sexual behavior of guinea pigs. *J. Pers. Soc. Psychol.* **2**, 460–464.

Gerall, H. D., Ward, I., and Gerall, A. A. 1967. Disruption of the male rat's sexual behaviour induced by social isolation. *Anim. Behav.* **15**, 54–58.

Gilbert, B. K. 1968. Development of social behavior in the fallow deer (*Dama dama*). *Z. Tierpsychol.* **25**, 867–876.

Ginsburg, B. E., and Jumonville, J. E. 1967. Genetic variability in response to early stimulation viewed as an adaptive mechanism in population ecology. *Amer. Zool.* **7**, 795.

Goodall, J. 1965. Chimpanzees of the Gombe stream reserve. *In* "Primate Behavior" (I. De Vore, ed.), pp. 425–473. Holt, Rinehart & Winston, New York.

Green, P. C. 1965. Influence of early experience and age on expression of affect in monkeys. *J. Genet. Psychol.* **106**, 157–171.

Grimm, R. J. 1967. Catalogue of sounds of the pigtailed macaque (*Macaca neimestrina*). *J. Zool.* **152**, 361–373.

Grosvenor, C. E. 1956. Some effects of ergotamine tartrate upon lactation in the rat. *Amer. J. Physiol.* **186**, 211–215.

Gruendel, A. D., and Arnold, W. J. 1969. Effects of early social deprivation on reproductive behavior of male rats. *J. Comp. Physiol. Psychol.* **67**, 123–128.

Guggisberg, C. A. W. 1961. "Simba. The Life of the Lion." Bailey Bros. & Swinfen, London.

Hård, E., and Larsson, K. 1968. Dependence of adult mating behavior in male rats on the presence of littermates in infancy. *Brain, Behav. Evol.* **1**, 405–419.

Hafez, E. S. E., and Schein, M. W. 1962. The behaviour of cattle. *In* "The Behaviour of Domestic Animals" (E. S. E. Hafez, ed.), pp. 347–369. Williams & Wilkins, Baltimore, Maryland.

Hafez, E. S. E., Williams, M., and Wierzbowski, S. 1962. The behaviour of horses. *In* "The Behaviour of Domestic Animals" (E. S. E. Hafez, ed.), pp. 370–396. Williams & Wilkins, Baltimore, Maryland.

Hall, K. R. L., and Mayer, B. 1967. Social interactions in a group of captive patas monkeys (*Erythrocebus patas*). *Folia Primatol.* **5**, 213–236.

Hamburg, D. A. 1969. Observations of mother-infant interactions in primate field studies. *In* "Determinants of Infant Behaviour" (B. M. Foss, ed.), Vol. IV, pp. 3–14. Methuen, London.

Hansen, E. W. 1966. The development of maternal and infant behaviour in the rhesus monkey. *Behaviour* **27**, 107–149.

Harlow, H. F. 1961. The development of affectional patterns in infant monkeys. *In* "Determinants of Infant Behaviour" (B. M. Foss, ed.), Vol. I. Wiley, New York.

Harlow, H. F. 1962. Development of affection in primates. *In* "Roots of Behavior" (E. L. Bliss, ed.), pp. 157–166. Harper & Row (Hoeber), New York.

Harlow, H. F., and Harlow, M. K. 1965. The affectional systems. *In* "Behavior of Non-human Primates" (A. M. Schrier, H. F. Harlow, and F. Stollnitz, eds.), Vol. II, pp. 287–334. Academic Press, New York.

Harlow, H. F., and Harlow, M. K. 1966. Learning to love. *Amer. Sci.* **54**, 244–270.

Harlow, H. F., and Harlow, M. K. 1969. Effects of various mother-infant relationships on rhesus monkey behaviors. *In* "Determinants of Infant Behaviour" (B. M. Foss, ed.), Vol. IV, pp. 15–36. Methuen, London.

Harlow, H. F., Harlow, M. K., Dodsworth, R. O., and Arling, G. L. 1966. Maternal behavior of rhesus monkeys deprived of mothering and peer associations in infancy. *Proc. Amer. Phil. Soc.* **110**, 58–66.

Harper, L. V. 1968a. The effects of isolation from birth on the social behaviour of guinea pigs at adulthood. *Anim. Behav.* **16**, 58–64.

Harper, L. V. 1968b. Isolated guinea pigs' responses to moving models. *Amer. Zool.* **8**, 746.

Harrisson, B. 1960. A study of orangutan behaviour in a semi-wild state 1956–1960. *Sarawak Mus. J.* **9**, 422–447.

Hediger, H. 1955. "Studies of the Psychology and Behaviour of Animals in Zoos and Circuses." Butterworth, London and Washington, D.C.

Hediger, H. 1965. Environmental factors influencing the reproduction of zoo animals. *In* "Sex and Behavior" (F. A. Beach, ed.), pp. 319–354. Wiley, New York.

Hersher, L., Richmond, J. B., and Moore, A. U. 1963. Maternal behavior in sheep and goats. *In* "Maternal Behavior in Mammals" (H. L. Rheingold, ed.), pp. 203–232. Wiley, New York.

Hinde, R. A., and Spencer-Booth, Y. 1967a. The effects of social companions on mother-infant relations in rhesus monkeys. *In* "Primate Ethology" (D. Morris, ed.), pp. 267–286. Aldine, Chicago, Illinois.

Hinde, R. A., and Spencer-Booth, Y. 1967b. The behaviour of socially living rhesus monkeys in their first two and a half years. *Anim. Behav.* **15**, 169–196.

Hinde, R. A., Spencer-Booth, Y., and Bruce, M. 1966. Effects of 6-day maternal deprivation on rhesus monkey infants. *Nature (London)* **210**, 1021–1023.

Hinde, R. A., Spencer-Booth, Y., and Bruce, M. 1966. Effects of 6-day maternal deprivation on rhesus monkey infants. *Nature (London)* **210**, 1021–1023.

Howard, W. E. 1949. Dispersal, amount of inbreeding and longevity in a local population of prairie deermice on the George reserve, southern Michigan. *Lab. Vert. Biol., Univ. Mich.* **43**, 1–50.

Hunter, R. F., and Davies, G. E. 1963. The effects of method of rearing on the social behavior of Scottish blackface hoggets. *Anim. Prod.* **5**, 183–194.

Itani, J. 1958. On the acquisition and propagation of a new food habit in the natural group of the Japanese monkey at Takasakiyama. *Primates* **1**, 84–86.

Itani, J. 1959. Paternal care in the wild Japanese monkey *Macaca fuscata fuscata. Primates* **2**, 61–93.

Jay, P. 1963. Mother-infant relations in langurs. *In* "Maternal Behavior in Mammals" (H. L. Rheingold, ed.), pp. 282–304. Wiley, New York.

Jensen, G. D., and Bobbitt, R. A. 1967. Implications of primate research for understanding infant development. *In* "The Exceptional Infant" (J. Hellmuth, ed.), Vol. 1. Special Child Publ., Seattle, Washington.

Jewell, P. A., and Loizos, C., eds. 1966. "Play, Exploration and Territory in Mammals," Symp. Zool. Soc. London No. 18. Academic Press, New York.

Jolly, A. 1966a. "Lemur Behavior." Univ. of Chicago Press, Chicago, Illinois.

Jolly, A. 1966b. Lemur social behavior and primate intelligence. *Science* **153**, 501–506.

Joslin, P. W. B. 1967. Movements and home sites of timber wolves in Algonquin park. *Amer. Zool.* **7**, 279–288.

Kabat, C., Collias, N. E., and Guettinger, R. C. 1953. Some Winter Habits of White Tailed

Deer and the Development of Census Methods in the Flag Yard of Northern Wisconsin. Wisconsin Conservation Department Technical Wildlife Bull. No. 7, Madison, Wisconsin.

Kaczmarski, F. 1966. Bioenergetics of pregnancy and lactation in the bank vole. *Acta Theriol.* **11**, 409–417.

Kaufman, I. C., and Rosenblum, L. A. 1967. The reaction to separation in infant monkeys: Anaclitic depression and conservation-withdrawal. *Psychosom. Med.* **29**, 648–675.

Kaufmann, J. H. 1962. Ecology and social behavior of the coati, *Nasua narica,* on Barro Colorado Island, Panama. *Univ. Calif., Berkeley, Publ. Zool.* **60**, 95–222.

Kaufmann, J. H. 1966. Behavior of infant rhesus monkeys and their mothers in a free ranging band. *Zoologica (New York)* **51**, 17–28.

Kawai, M. 1963. On the newly-acquired behaviors of the natural troop of Japanese monkeys on Koshima Island. *Primates* **4**, 113–115.

Kawai, M. 1965. Newly-acquired pre-cultural behavior of the natural troop of Japanese monkeys on Koshima islet. *Primates* **6**, 1–30.

Kawamura, S. 1959. The process of sub-culture propagation among Japanese macaques. *Primates* **2**, 43–54.

King, D. L. 1966. A review and interpretation of some aspects of the infant-mother relationship in mammals and birds. *Psychol. Bull.* **65**, 143–155.

King, J. A. 1955. Social behavior, social organization and population dynamics in a black-tailed prairiedog town in the Black Hills of South Dakota. *Contrib. Lab. Vert. Biol., Univ. Mich.* **67**, 1–128.

Koford, C. B. 1957. The vicuna and the puna. *Ecol. Monogr.* **27**, 153–219.

Kummer, H. 1967. Tripartite relations in Hamadryas baboons. *In* "Social Communication among Primates" (S. A. Altmann, ed.), pp. 63–71. Univ. of Chicago Press, Chicago, Illinois.

Kummer, H. 1968. "Social Organization of Hamadryas Baboons." Univ. of Chicago Press, Chicago, Illinois.

Kunkel, P., and Kunkel, J. 1964. Beitrage zur Ethologie des Hausmeerschweinchens *Cavia apera* f. porcellus (L). *Z. Tierpsychol.* **21**, 602–641.

Kuo, Z. Y. 1930. The genesis of the cat's response to the rat. *J. Comp. Psychol.* **11**, 1–35.

Kuo, Z. Y. 1938. Further study on the behavior of the cat toward the rat. *J. Comp. Psychol.* **25**, 1–8.

Lahiri, R. K., and Southwick, C. H. 1966. Parental care in *Macaca sylvana. Folia Primatol.* **4**, 257–264.

Lehrman, D. S. 1956. On the organization of maternal behavior and the problem of instinct. *In* "L'Instinct dans le Comportement des Animaux et de l'Homme" (P. -P. Grasse, ed.), pp. 475–513. Masson, Paris.

LeMangen, J., and Tallon, S. 1968. Preference alimentaire du jeune rat induite par l'allaitement maternel. *C. R. Soc. Biol.* **162**, 387–390.

Lemmon, W. B. 1968. Delivery and maternal behavior in captive reared primiparous chimpanzees, *Pan troglodytes. Amer. Zool.* **8**, 740.

Lent, P. C. 1966. Calving and related social behavior in the barren-ground caribou. *Z. Tierpsychol.* **23**, 701–756.

Levine, S., and Mullins, R. F. 1966. Hormonal influences on brain organization in infant rats. *Science* **152**, 1585–1592.

Leyhausen, P. 1968. Growth and maturation of the killing instinct in cats, both great and small. Lecture delivered at Smithsonian Institution, Washington, D.C.

Liddell, H. S. 1954. Conditioning and emotions. *Sci. Amer.* **160**, 48–57.

Liddell, H. S. 1960. Experimental neurosis in animals. *In* "Stress and Psychiatric Disorder" (J. M. Tanner, ed.). Blackwell, Oxford.

Linsdale, J. M., and Tomich, P. Q. 1953. "A Herd of Mule Deer." Univ. of California Press, Berkeley, California.

Lockley, R. M. 1961. Social structure and stress in the rabbit warren. *J. Anim. Ecol.* **30**, 385–423.

Lowe, V. P. W. 1966. Observation on the dispersal of red deer on Rhum. In "Play, Exploration and Territory in Mammals" (P. A. Jewell and C. Loizos, eds.), Symp. Zool. Soc. London No. 18, pp. 211–228. Academic Press, New York.

McCarley, H. 1966. Annual cycle, population dynamics and adaptive behavior of *Citellus tridecemlineatus. J. Mammal.* **47**, 294–316.

McHugh, T. 1958. Social behavior of the American buffalo (*Bison bison bison*). *Zoologica (New York)* **43**, 1–40.

Mainardi, D. 1963. Speciazone nel topo. *Inst. Lomb. (B)* **97**, 135–142.

Mainardi, D., Marsan, M., and Pasquali, A. 1965. Causation of sexual preferences of the house mouse. The behavior of mice reared by parents whose odor was artificially altered. *Atti Soc. Ital. Sci. Natur. Milan* **104**, 325–338.

Marr, J. N., and Gardner, L. E. 1965. Early olfactory experience and later social behavior in the rat: preference, sexual responsiveness and care of young. *J. Genet Psychol.* **107**, 167–174.

Marsden, H. M. 1968. Agonistic behaviour of young rhesus monkeys after changes induced in social rank of their mothers. *Anim. Behav.* **16**, 38–44.

Martinek, Z., and Lat, J. 1968. Ontogenetic differences in spontaneous reactions of dogs to new environment. *Physiol. Bohemoslov.* **17**, 545–552.

Martinsen, D. L. 1968. Temporal patterns in the home ranges of chipmunks (*Eutamis*). *J. Mammal.* **49**, 83–91.

Mason, W. A. 1963. The effects of environmental restriction on the social development of rhesus monkeys. In "Primate Social Behavior" (C. H. Southwick, ed.), pp. 161–173. Van Nostrand, New York.

Mason, W. A. 1965. Determinants of social behavior in young chimpanzees. In "Behavior of Nonhuman Primates" (A. Schrier, H. F. Harlow, and F. Stollnitz, eds.), Vol. II, pp. 335–363. Academic Press, New York.

Mason, W. A. 1968. Early social deprivation in the nonhuman primates: Implications for human behavior. In "Environmental Influences, Biology and Behavior Series" (D. C. Glass, ed.), pp. 70–101. Rockefeller Univ. Press, New York.

Mason, W. A., Harlow, H. F., and Rueping, R. R. 1959. The development of manipulatory responsiveness in the infant rhesus monkey. *J. Comp. Physiol. Psychol.* **52**, 555–558.

Masserman, J. H., Wechkin, S., and Woolf, M. 1968. Social relationships and aggression in rhesus monkeys. *Arch. Gen. Psychiat.* **18**, 210–213.

Mech, L. D. 1966. The Wolves of Isle Royale. U.S. National Parks Service Fauna Ser. No. 7. U.S. Govt. Printing Office, Washington, D.C.

Meier, G. W. 1965. Maternal behaviour of feral- and laboratory-reared monkeys following the surgical delivery of their infants. *Nature (London)* **206**, 492–493.

Meites, J. 1966. Control of mammary growth and lactation. In "Neuroendocrinology" (L. Martini and W. F. Ganong, eds.), Vol. 1. Academic Press, New York.

Melzack, R. 1968. Restricted Dogs. Film. National Film Board of Canada, Toronto.

Melzack, R., and Burns, S. K. 1965. Neurophysiological effects of early sensory restriction. *Exp. Neurol.* **13**, 163–175.

Menzel, E. W., Jr. 1964. Patterns of responsiveness in chimpanzees reared through infancy under conditions of environmental restriction. *Psychol. Forsch.* **27**, 337–365.

Menzel, E. W., Jr. 1966. Responsiveness to objects in free-ranging Japanese monkeys. *Behaviour* **26**, 130–150.

Menzel, E. W., Jr. 1967. Naturalistic and experimental research on primates. *Hum. Develop.* **10,** 170–186.

Miller, R. E., Caul, W. F., and Mirsky, I. A. 1967. Communication of affects between feral and socially isolated monkeys. *J. Pers. Soc. Psychol.* **7,** 231–239.

Mitchell, G. D. 1968. Persistent behavior pathology in rhesus monkeys following early social isolation. *Folia Primatol.* **8,** 132–147.

Mitchell, G. D., Ruppenthal, G. C., Raymond, E. J., and Harlow, H. F. 1966. Long-term effects of multiparous and primiparous monkey mother rearing. *Child Develop.* **37,** 781–792.

Mitchell, G. D., Arling, G. L., and Møller, G. W. 1967. Long-term effects of maternal punishment on the behavior of monkeys. *Psychonomic Sci.* **8,** 209–210.

Moltz, H., and Wiener, E. 1966. Effects of ovariectomy on maternal behavior of primiparous and multiparous rats. *J. Comp. Physiol. Psychol.* **62,** 382–387.

Moltz, H., Levin, R., and Leon, M. 1969a. Differential effects of progesterone on the maternal behavior of primiparous and multiparous rats. *J. Comp. Physiol. Psychol.* **67,** 36–40.

Moltz, H., Levin, R., and Leon, M. 1969b. Prolactin in the postpartum rat: Synthesis and release in the absence of suckling stimuli. *Science* **163,** 1083–1084.

Murie, A. 1944. The Wolves of Mount McKinley. U.S. National Park Service Fauna Ser. No. 5. U.S. Govt. Printing Office, Washington, D.C.

Nagy, Z. M. 1965. Effect of early environment upon later social preferences in two species of mice. *J. Comp. Physiol. Psychol.* **60,** 98–101.

Nelson, J. E. 1965. Behaviour of Australian pteropodidae (Megachiroptera). *Anim. Behav.* **13,** 544–557.

Nicoll, C. S., and Meites, J. 1959. Prolongation of lactation in the rat by litter replacement. *Proc. Soc. Exp. Biol. Med.* **101,** 81–82.

Nissen, H. W. 1956. Development of Sexual Behavior in Chimpanzees. Unpublished manuscript cited in Young, 1957, p. 91.

Noirot, E. 1964. Changes in responsiveness to young in the adult mouse. I. The problematical effect of hormones. *Anim. Behav.* **12,** 52–58.

Ota, K., and Yokoyama, A. 1967a. Body weight and food consumption of lactating rats: effects of ovariectomy and of arrest and resumption of suckling. *J. Endocrinol.* **38,** 251–261.

Ota, K., and Yokoyama, A. 1967b. Body weight and food consumption of lactating rats nursing various sizes of litters. *J. Endocrinol.* **38,** 263–268.

Pearson, O. P. 1944. Reproduction in the shrew (*Blarina brevicauda* Say). *Amer. J. Anat.* **75,** 39–93.

Ploog, D. W. 1967. The behavior of squirrel monkeys (*Saimiri sciureus*) as revealed by sociometry, bioacoustics and brain stimulation. *In* "Social Communication among Primates" (S. A. Altman, ed.), pp. 149–184. Univ. of Chicago Press, Chicago, Illinois.

Quadagno, D. M., and Banks, E. M. 1968. The effect of reciprocal cross-fostering on the social preference, exploratory, sexual and aggressive behavior of the pigmy mouse, *Baiomys taylori,* and the house mouse, *Mus musculus,* C 57 Br/J. *Amer. Zool.* **8,** 746–747.

Rabb, G. B., Woolpy, J. H., and Ginsburg, B. E. 1967. Social relationships in a group of captive wolves. *Amer. Zool.* **7,** 305–311.

Ray, C. 1967. Natural history of the Weddell seal in Antarctica. *Amer. Zool.* **7,** 808.

Rheingold, H. L. 1963a. Introduction. *In* "Maternal Behavior in Mammals" (H. L. Rheingold, ed.), pp. 1–7. Wiley, New York.

Rheingold, H. L., ed. 1963b. "Maternal Behavior in Mammals." Wiley, New York.

Richards, M. P. M. 1967. Maternal behaviour in rodents and lagomorphs. *In* "Advances in Reproductive Physiology" (A. McLaren, ed.), Vol. 2. Logos Press, London.

Ripley, S. 1967. Intertroop encounters among ceylon gray langurs (*Presbytis entellus*). *In* "Social Communication among Primates" (S. A. Altmann, ed.), pp. 237–253. Univ. of Chicago Press, Chicago, Illinois.

Roberts, W. W., and Bergquist, E. H. 1968. Attack elicited by hypothalamic stimulation in cats raised in social isolation. *J. Comp. Physiol. Psychol.* **66**, 590–595.

Romer, A. S. 1958. Phylogeny and behavior with special reference to vertibrate evolution. *In* "Behavior and Evolution" (A. Roe and G. G. Simpson, eds.), pp. 48–75. Yale Univ. Press, New Haven, Connecticut.

Romer, A. S. 1967. Major steps in vertibrate evolution. *Science* **158**, 1629–1637.

Rosenblatt, J. S. 1967. Nonhormonal basis of maternal behavior in the rat. *Science* **156**, 1512–1514.

Rosenblatt, J. S., and Schneirla, T. C. 1962. The behavior of cats. *In* "The Behaviour of Domestic Animals" (E. S. E. Hafez, ed.), pp. 453–488. Williams & Wilkins, Baltimore, Maryland.

Rosenblatt, J. S., Turkewitz, G., and Schneirla, T. C. 1961. Early socialization in the domestic cat as based on feeding and other relationships between female and young. *In* "Determinants of Infant Behaviour" (B. M. Foss, ed.), pp. 51–74. Wiley, New York.

Rosenblum, L. A. 1968. Mother-infant relations and early behavioral development in the squirrel monkey. *In* "The Squirrel Monkey" (L. A. Rosenblum and R. W. Cooper, eds.). Academic Press, New York.

Rothchild, I. 1967. The neurological basis for the anovulation of the luteal phase, lactation and pregnancy. *In* "Reproduction in the Female Mammal" (G. E. Lamming and E. C. Amoroso, eds.). Plenum Press, New York.

Rowell, T. E., Hinde, R. A., and Spencer-Booth, Y. 1964. "Aunt"-infant interaction in captive rhesus monkeys. *Anim. Behav.* **12**, 219–226.

Ruffer, D. G. 1965. Sexual behaviour of the northern grasshopper mouse (*Onychomys leucogaster*). *Anim. Behav.* **13**, 447–452.

Sackett, G. P. 1966. Monkeys reared in isolation with visual input: Evidence for an innate releasing mechanism. *Science* **154**, 1468–1473.

Sackett, G. P. 1967. Some persistent effects of different rearing conditions on preadult social behavior of monkeys. *J. Comp. Physiol. Psychol.* **64**, 363–365.

Sackett, G. P., Porter, M., and Holmes, H. 1965. Choice behavior in rhesus monkeys: Effects of stimulation during the first month of life. *Science* **147**, 304–306.

Sade, D. S. 1967. Determinants of dominance in a group of free-ranging rhesus monkeys. *In* "Social Communication among Primates" (S. A. Altmann, ed.), pp. 99–114. Univ. of Chicago Press, Chicago, Illinois.

Salzen, E. A. 1967. Imprinting in birds and primates. *Behaviour* **28**, 232–254.

Sauer, E. G. F. 1967. Mother-infant relationship in galagos and the oral child-transport among primates. *Folia Primatol.* **7**, 127–149.

Schaller, G. B. 1963. "The Mountain Gorilla." Univ. of Chicago Press, Chicago, Illinois.

Schaller, G. B. 1967. "The Deer and the Tiger." Univ. of Chicago Press, Chicago, Illinois.

Scheffer, V. B. 1950. Experiments in the Marking of Seals and Sea Lions. *U.S. Fish Wildl. Serv., Spec. Sci. Rep. Wildl.* **4**.

Schenkel, R. 1966a. On sociology and behaviour in impala (*Aepyceros melampus* Lichenstein). *East Afr. Wildl. J.* **4**, 99–114.

Schenkel, R. 1966b. Play, exploration and territoriality in the wild lion. *In* "Play, Exploration and Territory in Mammals" (P. A. Jewell and C. Loizos, eds.), Symp. Zool. London No. 18, pp. 11–22. Academic Press, New York.

Schneirla, T. C., and Rosenblatt, J. S. 1961. Behavioral organization and genesis of the social bond in insects and mammals. *Amer. J. Orthopsychiat.* **31**, 223–253.

Schneirla, T. C., Rosenblatt, J. S., and Tobach, E. 1963. Maternal behavior in the cat. *In* "Maternal Behavior in Mammals" (H. L. Rheingold, ed.), pp. 122–168. Wiley, New York.

Scott, J. P. 1962. Critical periods in behavioral development. *Science* **138**, 949–958.

Scott, J. P. 1967. The development of social motivation. *In* "Nebraska Symposium on Motivation" (D. Levine, ed.), pp. 111–132. Univ. of Nebraska Press, Lincoln, Nebraska.

Scott, J. P., and Fuller, J. L. 1965. "Genetics and the Social Behavior of the Dog." Univ. of Chicago Press, Chicago, Illinois.

Seay, B. 1966. Maternal behavior in primiparous and multiparous monkeys. *Folia Primatol.* **4**, 146–168.

Seay, B., and Harlow, H. F. 1965. Maternal separation in the rhesus monkey. *J. Nerv. Ment. Dis.* **140**, 434–441.

Seay, B., Hansen, E. W., and Harlow, H. F. 1962. Mother-infant separation in monkeys. *J. Child Psychol. Psychiat.* **3**, 123–132.

Seay, B., Alexander, B. K., and Harlow, H. F. 1964. Maternal behavior of socially deprived rhesus monkeys. *J. Abnorm. Soc. Psychol.* **69**, 345–354.

Selye, H., and McKeown, T. 1934. Further studies on the influence of suckling. *Anat. Rec.* **60**, 323–332.

Senko, M. G. 1966. Effects of early, intermediate and late experiences upon adult macaque sexual behavior. Unpublished MA. Thesis Univ. of Wisconsin, Madison, Wisconsin.

Sharp, W. M. 1959. A commentary on the behavior of free-running gray squirrels. *In* "Symposium on the Gray Squirrel" (V. Flyger, ed.). Md. Dep. of Res. and Educ., Annapolis, Maryland.

Sharp, W. M., and Sharp, L. H. 1956. Nocturnal movements and behavior of wild raccoons at a winter feeding station. *J. Mammal.* **37**, 170–177.

Simpson, G. G. 1950. "The Meaning of Evolution." Yale Univ. Press, New Haven, Connecticut.

Sluckin, W. 1965. "Imprinting and Early Learning." Aldine, Chicago, Illinois.

Smith, C. C. 1968. The adaptive nature of social organization in the genus of tree squirrels *Tamaisciureus. Ecol. Monogr.* **38**, 31–63.

Sorenson, M. W., and Conaway, C. H. 1966. Observations on the social behavior of tree shrews in captivity. *Folia Primatol.* **4**, 124–145.

Southern, H. N. 1940. The ecology and population dynamics of the wild rabbit (*Oryctolagus cuniculus*). *Ann. Appl. Biol.* **27**, 509–526.

Spencer-Booth, Y., and Hinde, R. A. 1967. The effects of separating rhesus monkey infants from their mothers for six days. *J. Child Psychol. Psychiat.* **7**, 179–198.

Stephenson, G. R. 1967. Cultural acquisition of a specific learned response among rhesus monkeys. *In* "Neue Ergebnisse der Primatologie" (D. Starck, R. Schneider, and H. J. Kuhn, eds.), pp. 279–288. Fischer, Stuttgart.

Sugiyama, Y. 1960. On the division of a natural troop of Japanese monkeys at Takasakiyama. *Primates* **2**, 109–148.

Suzuki, A. 1965. An ecological study of wild Japanese monkeys in snowy areas focused on their food habits. *Primates* **6**, 31–72.

Tavolga, M. C., and Essapian, F. S. 1957. The behavior of the bottle-nosed dolphin (*Tursiops truncatus*): Mating, pregnancy, parturition and mother-infant behavior. *Zoologica (New York)* **42**, 11–31.

Tevis, L. 1950. Summer behavior of a family of beavers in New York State. *J. Mammal.* **31**, 40–65.

Thoman, E. B., and Arnold, W. J. 1968a. Effects of incubator rearing with social deprivation on maternal behavior in rats. *J. Comp. Physiol. Psychol.* **65**, 441–446.

Thoman, E. B., and Arnold, W. J. 1968b. Incubator rearing of infant rats without the mother: Effects on adult emotionality and learning. *Develop. Psychobiol.* **1**, 219–222.

Thorpe, W. H. 1962. "Learning and Instinct in Animals." Harvard Univ. Press, Cambridge, Massachusetts.

Thorpe, W. H. 1966. Ritualization in ontogeny: I. Animal play. *Phil. Trans. Roy. Soc. London, Ser. B* **251**, 311–319.

Tinbergen, N. 1960. Behaviour, systematics and natural selection. *In* "Evolution after Darwin. Vol. I. The Evolution of Life" (S. Tax, ed.), pp. 595–613. Univ. of Chicago Press, Chicago, Illinois.

Tsumori, A. 1967. Newly acquired behavior and social interactions of Japanese monkeys. *In* "Social Communication among Primates" (S. A. Altmann, ed.), pp. 207–219. Univ. of Chicago Press, Chicago, Illinois.

Vandenbergh, J. G., and Vessey, S. 1966. Seasonal breeding in free-ranging rhesus monkeys and related ecological factors. *Amer. Zool.* **6**, 342.

van Lawick-Goodall, J. 1967. Mother-offspring relationships in free-ranging chimpanzees. *In* "Primate Ethology" (D. Morris, ed.), pp. 287–346. Aldine, Chicago, Illinois.

Washburn, S. L., and Hamburg, D. A. 1965. The study of primate behavior. *In* "Primate Behavior" (I. De Vore, ed.), pp. 607–622. Holt, Rinehart & Winston, New York.

Wecker, S. C. 1963. The role of early experience in habitat selection by the prairie deer mouse, *Peromyscus maniculatus bairdi. Ecol. Monogr.* **33**, 307–325.

Welker, W. I. 1956. Some determinants of play and exploration in chimpanzees. *J. Comp. Physiol. Psychol.* **49**, 84–89.

Welker, W. I. 1959. Genesis of exploratory and play behavior in infant raccoons. *Psychol. Rep.* **5**, 764.

Woolpy, J. H., and Ginsburg, B. E. 1967. Wolf socialization: A study of temperament in a wild social species. *Amer. Zool.* **7**, 357–363.

Wright, P. L. 1948. Breeding habits of captive long-tailed weasels *(Mustela frenata). Amer. Midl. Natur.* **39**, 338–343.

Wynne-Edwards, V. C. 1962. "Animal Dispersion in Relation to Social Behaviour." Oliver & Boyd, Edinburgh and London.

Young, W. C. 1957. Genetic and psychological determinants of sexual behavior patterns. *In* "Hormones, Brain Function and Behavior" (H. Hoagland, ed.), pp. 75–98. Academic Press, New York.

Zarrow, M. X., Denenberg, V. H., and Anderson, C. O. 1965. Rabbit: Frequency of suckling in the pup. *Science* **150**, 1835.

Zimbardo, P. G. 1958. The effects of early avoidance training and rearing conditions upon the sexual behavior of the male rat. *J. Comp. Physiol. Psychol.* **51**, 764–769.

The Relationships between Mammalian Young and Conspecifics Other Than Mothers and Peers: A Review

Y. Spencer-Booth

SUB-DEPARTMENT OF ANIMAL BEHAVIOUR
MADINGLEY, CAMBRIDGE, ENGLAND

I. INTRODUCTION

It is characteristic of mammals that the females show care for their young, and such behavior, including activities such as nest-building before and after parturition, licking, retrieving, and suckling the young are generally grouped under the umbrella term "maternal behavior." The category is delimited by function, namely care for the young, but the type of care involved is very different in the different orders of mammals. The young may be deposited in a specially prepared nest, where the mother may spend much of her time, or which she may visit only briefly once in 48 hours. If no nest is made, the young may be carried continuously by the mother for some time after birth, or they may be left hidden in an unprepared site. The degree of helplessness of the young at birth varies widely, and so of course does the time for which care is required.

Not only is the term "maternal behavior" applied to descriptively different types of behavior, but the meaningfulness of the concept of "maternal behavior" applied even within one species has been challenged a number of times on the grounds that the different behavior patterns included are not well correlated (e.g., Denenberg et al., 1958; Rowell, 1960; Herscher et al., 1963a); it is thus not a causal category although common causal factors are often implied.

However, the broad use of the term in a functional capacity may be useful in some contexts, provided its limitations are realized. But then an additional complication arises if the behavior that the mother shows to her young is also shown by conspecifics other than the true mother. In some species considerable care for the young is shown by the male parent. However, the term "parental behavior" is unsatisfactory (see also Lehrman, 1961), as not only may the male show different patterns of care from those shown by the mother, but other male or female group companions, including females which have not borne young themselves, sometimes show at least recognizable elements of the patterns of care shown by the mother. This is true of some species in natural conditions, although in some only in rather artificial ones. It is especially characteristic of primates but is found in some other orders including the carnivores and rodents.

The above observations are important when the control of any aspect of the mother's behavior to her young comes under investigation. One should know what there is about a mother's behavior that is peculiar, within that species, to a female interacting with her own young before investigating what it is about the female's internal milieu that is affecting her behavior.

Another important aspect of this network of problems is the question of how specific females are in their responsiveness to young. Thus, given that a female is in the physiological and experiential state in which one would expect "maternal" responses to be shown, to what range of objects will she show them; for instance, do they have to be the young which she has borne, or can they be the young of other females of the same or even of different species? This leads to the question of the formation of social attachments between mothers and their young. Furthermore, much "maternal behavior" can be recognized only in the presence of young (for instance, suckling) and depends on their co-operation, and it is thus important to know how specific the young are to their own mothers.

Before anything can be said about the possible mechanisms involved in the formation of individual mother-infant relationships, and in particular before any generalizations can be made, it is important to be clear about the precise conditions under which they form and are maintained. The relationship depends on the behavior of both mother and young, and thus the opportunities for interactions of mothers with young other than their own, and of young with all members of the social group in which they are reared, other than their own mothers, must be taken into account.

With these problems in mind this review sets out to survey the field on a broad basis, to show to what extent within the mammals members of the species other than the natural mother are known to show interactions with the young, and to what extent the behavior duplicates that of the

mother. Further, if only some members of the species show such behavior, how is this related to the extent of their previous experience with
young and also to their sex and demographic category? Finally how
specific are lactating females in their responsiveness, and how specific
are the young, and under what conditions is such specificity acquired
and maintained?

The information, in some cases extremely scrappy, comes from a wide
range of conditions, ranging from animals in the wild, through semi-
captivity to domesticity and close confinement in zoos, and from experimental animals in laboratories. These varied conditions under which
the animals were living must be taken into account in surveying the data.
In general, interactions between conspecifics only are discussed.

II. Behavior of Males Toward Young

Male mammals show a wide range of types of behavior to young of the
species, ranging from outright aggression and killing them, to intense
care. In species in which the males are prone to killing young, the females normally prevent them from having access to their offspring;
there are also species in which the males normally do not have access
to the young, yet will not necessarily attack them, and may even show
care for them, in abnormal circumstances. The care shown by males may
be rather similar to that shown by the mother or other females, although
males never lactate unless severely interfered with experimentally (e.g.,
Nelson and Smelser, 1933); in most cases, however, careful comparisons
have not been made. In some species the males spend more time caring
for the young than the females, and in others the male plays an important role in sharing in the provision of food.

The information is arranged according to the orders; for some orders
there seem to be no data.

A. Marsupials

There seems to be virtually no information for marsupials, but a male
ringtail possum *(Pseudocheirus peregrinus)* has been seen to help the female to nest-build both in and out of the breeding season, to accompany
the female when carrying pouch young, and to be left alone with post-
pouch young (Thompson and Owen, 1964).

B. Insectivora

There is very little information for the Insectivora, but Conoway
(1958) observed captive groups of the least shrew *(Cryptotis parva)* and
noted that males as well as other females showed considerable "maternal

behavior" but did not specify what kind. He implied that when a litter was scattered the males participated in caring for the young and re-building the nest, and that they ran about the cage carrying young. The litter was successfully reared. Gould and Eisenberg (1966) observed in a captive tenrec pair *(Hemicentites semispinosus)* that the litter huddled round the male while the female was nest-building, and that the male as well as the female pressed his snout against a baby wandering from the nest. However, unlike the mother, the male was not observed to pick up the young to retrieve them.

Southern (1964) says that in the hedgehog *(Erinaceous europaeus)* the male takes no part in rearing the young.

C. CHIROPTERA

Among the Chiroptera the males are often excluded from the nursing colonies into which the young are born (e.g., British horse-shoe bats, Matthews, 1937; *Corynorhinus rafinesqui,* Pearson *et al.,* 1952; *Anthrozous pallidus,* Beck and Rudd, 1960) although an occasional male, often juvenile, may be found in such colonies. However, Davis *et al.* (1968) found some adult males in nearly all colonies of *Eptesicus fuscus* ex-amined, Rice (1957) found a few males in the maternity colonies of *Myotis austroriparius,* and Orr (1954) found both sexes in varying pro-portions in roosts of *Antrozous pallidus* in which there were young.

Causey and Waters (1936) mention a description (Gill, Standard Natural History) of *Cheiromeles torquatus* males as well as females having a nursing pouch, and that when the female has twins the male may re-lieve her of the charge of one of them, carrying it in the pouch. However, other authors think that the extremities of the folded wings are put into these pouches (e.g., Morris, 1965; Brosset, 1966).

Kulzer (1958) noted that in captive *Rousetus* sp. the male enveloped the young in his wing membrane and groomed it, commenting that his behavior resembled that shown to the female during mating.

D. PRIMATES

Among the primates one can find the full range of male behavior, from aggression and killing to intense care.

1. Prosimii

At least some species of *Tupaia* deposit their young in nests, and one of the most fascinating instances of male behavior was observed in captive *Tupaia* (species not specified), in which the males built separate maternal nests unaided, 1–5 days before the females gave birth (Martin, 1966).

The intriguing problem is how this behavior is coordinated in time with parturition, and it would be well worth further investigation. However, the males showed no further care for the young, and indeed the females visited the nests only to feed the infants for 5 minutes once every 48 hours after parturition and showed no further care or concern for them. Sorenson and Conaway (1966) found that a litter of two *Tupaia tana* reared in the presence of two adult males began to be molested by them when nearly 3 months old, and were actively protected by the mother. However, by this time they would have been approaching sexual maturity (at four months).

Petter (1965) says that the Indriidae young are attractive to other group members including the male, and that *Lemur macaco* males frequently lick or play with young infants. The evidence for *Galago senegalensis moholi*, the lesser bush baby, seems rather conflicting. Doyle *et al.* (1967), studying animals under seminatural conditions, found that a male which was sharing a cage with a female which had twins showed no interest in the babies. However, Lowther (1940) kept the male separate from the female for three weeks after the birth of twins, and during this time he showed much interest in them and tried to get into their cage. When allowed in, he seemed solicitous of the young, but no details of his behavior are given. Sauer (1967) isolated a female *G. crassicaudatus monteira* at parturition and noted extreme interest in her babies from her previous companions including males, which were in an adjacent cage.

2. Anthropoidea

In the marmosets and tamarins the males seem to play an important role in caring for the young (Sanderson, 1957), spending more time caring for them than does the female. Most of the information relates to the common marmoset, *Hapale (= Callithrix) jacchus,* in captivity. Langford (1963) said that if there is any difficulty in expelling the fetus the male will assist with his hands. Lucas *et al.* (1937) also observed a male assisting at parturition. Both papers refer to the male receiving and cleaning the babies, and Fitzgerald (1935) found the male parent hugging and licking a stillborn infant the day after birth. After parturition the male normally carries the young, handing them back to the mother for nursing (Lucas *et al.,* 1927; Langford, 1963). Stellar (1960) said that the male continues to carry the young after weaning. Langford also observed the male premasticating solid food for infants about a month old. Hill (1957) noted the same behavior for *Hapale (= Callithrix) penicillata* and also observed that the young returned to the male if alarmed.

Much the same pattern seems to be shown by the tamarins *Oepidomidas* (= *Saguinus*) *oedipus,* in which Hampton (1964) noted that it is the male who carries the young almost all the time and responds to their distress calls. Hampton *et al.* (1966) observed that when two adult males were present in a group in which there was an infant, both males carried it; although they thought it possible that one was more active in doing so than the other, they were unable to say which was the father on this basis. Ditmars (1933) reported that in *Leontidus* sp. the young clung to the mother for the first week and were then carried by the male and returned to the female only at feeding time.

Moynihan (1964) found that of three infant night monkeys (*Aotus trivirgatus*) on Barro Colorado island, two were being carried by the male parent for most of the time that they were not being suckled, and the third infant was certainly carried by the male for much of the time. Moynihan (1964) said that other New World primates, such as the howler monkeys, *(Alouatta)* and the spider monkeys *(Ateles),* are carried only by their mothers, but Carpenter (1934) said that although males sometimes act aggressively to infants they also may protect them, for instance from human observers, and Carpenter (1965) noted that male *Alouatta palliata* gathered round an infant that had fallen from a tree. Wagner (1956) reported that the female spider monkeys leave the group shortly before giving birth, and return 2–4 months later. Williams (1967) reported that in a captive group of woolly monkeys *(Lagothrix lagotricha)* a male watched a female closely during parturition, stayed near her for 2 hours after the birth, and for several days afterward chased away other monkeys who seemed to be annoying the mother.

Mason is quoted by Napier and Napier (1967) as saying that, in the wild, the *Callicebus moloch* male carries the young except when it is being fed by its mother. Hanif (1967) observed a pair of zoo-living white-headed saki monkeys *(Pithecia pithecia)* and noted that the male made several unsuccessful attempts to take the baby, which was never seen to leave the mother for nearly 4 months. It may be that under more natural conditions the male would carry the baby.

In the squirrel monkey *(Saimiri sciureus),* Bowden, Winter, and Ploog (1967) noted that adult males caged with a parturient female retreated to the far end of the cage when the fetus began to appear, but Hopf (1967) who also worked with laboratory-living animals found that although interactions between infants up to a year old and adult males were rare, an adult male would defend the infant against group members that handled it too roughly.

In the Cercopithecidae the males do not usually carry the infants much, although they may show interest in them. van Kirchshofer (1960) noted that a zoo-living colobus monkey male *(Colobus polykomos kirkuyuinsis)* treated a baby kindly, and *Zoonooz* (1964) reported a baby as being handled occasionally by the male. Similarly in a zoo-living proboscis monkey *(Nasalis larvatus orientalis)* pair which had an infant, the male was said to be very tolerant (Pournelle, 1967). However an infant talapoin *(Miopithecus talapoin)* born to laboratory-living parents was avoided by the male, who jumped away if it touched him (Hill, 1966). An adult male patas monkey *(Erythrocebus patas)* living in captivity showed no interest in an infant born into the group (Hall and Mayer, 1967). Jay (1962) reported that in wild-living groups of langurs *(Presbytis entellus)* the adult male showed no interest in newborn or small infants, although C. McCann (1934) implies that an adult male tried to rescue a baby marooned in a tree after its mother had been shot. Sugiyama (1965a) described drastic behavior by males of this species, when taking over the leadership of groups in the wild. He reported that all the infants in the troops were bitten to death by the new leaders and thought that this might strengthen the ties between the male and the adult females of a troop. A male *Presbytis obscura* in a zoo-lving group was reported to show great interest in the babies, but not to be allowed to carry them (Badham, 1967a).

Among the macaques, *Macaca mulatta* males in captive groups show little interest in infants, although they do show some, especially when young and also in abnormal circumstances, such as when the infant's mother has been removed from the group (Spencer-Booth, 1968). The adult male may then cuddle the infant as it clings ventroventrally to him. In the wild, males of this species were seen, on rare occasions only, to associate closely with an infant for several hours by Southwick *et al.* (1965), but Kaufmann (1966) never saw mature males approach infants. However, in wild *M. fuscata* Itani (1959) found that in some troops there were relationships between particular male-infant pairs: the male was reported to hug and carry the infant. Such behavior apparently helps a male to establish himself in the center of a troop (Itani, 1961). MacRoberts is quoted by Washburn *et al.* (1965) as saying that there was a very high frequency of male care in Gibraltar macaques *(M. sylvana)*. Deag (1969) has also observed this in wild-living groups in Morocco. Males would carry babies and apparently use them as a kind of "passport" to approach other males, and might seek out a baby during or after agonistic interactions in the group.

Booth (1962) found that in *Cercopithecus aethiops, C. neglectus,* and *C. nitis* the males would pick up an infant which had been left alone.

Struhsaker (1967a) observed wild *C. aethiops* and only once saw an infant cling to an adult male, although juvenile males did handle infants.

"Parental" care seems to be more frequent in baboons than in other members of the Cercopithecidae. DeVore (1963) and Hall and DeVore (1965) observed free-ranging baboons *(Papio ursinus* and *P. anubis)* and found that juvenile and young adult males showed only perfunctory interest in the infants, while adult males frequently touched them, would carry them ventroventrally, and protect them after the mother had ceased to do so. An infant which had lost its mother was "adopted" by a male. Bolwig (1958) said that whereas some *P. ursinus* males were completely intolerant of infants, others showed great interest in them, and he saw two males with infants which they had adopted.

Kummer (1967, 1968) has studied *P. hamadryas* in the wild and in zoos. He describes adult males carrying young infants ventroventrally during troop movements, while the mothers walked alongside. Furthermore subadult or young adult males became the center of play groups, and acted as protector for any infant that was frightened, by threatening the aggressor. Kummer pointed out that these relationships were temporary ones, as an infant that was protected one moment might, if it attacked another infant, be threatened subsequently. Kummer thinks that the female infants transfer the mother's role to the young male, as they later form a one-male group around him. However, it may be that the initiative in making this transfer is largely taken by the male, as Kummer describes how subadult males kidnapped young infants, cuddling and carrying them for periods up to half an hour, and that motherless infants were invariably adopted by such males. In both cases the infants might attempt to escape from their kidnappers. These patterns abruptly vanished once the males had their own females.

Among the apes too the males show interest in and tolerance for the infants. Carpenter (1940) found that in wild gibbons *(Hylobates lar)* the very young infants were inspected and groomed by adult males. This author reports that in a captive group of *H. agilis* a small juvenile was carried for much of the day by a male. However, a captive male pileated gibbon *(Hylobates lar pileatus)* was interested in the baby, but was never allowed to carry it (Badham, 1967b). An orangutan male *(Pongo pygmaeus)* played with a baby which often went to it (Chaffee, 1967). Schaller (1965) describes wild male gorillas *(Gorilla gorilla)* as carrying infants short distances, but implies that this was initiated by the infants and merely tolerated by the males. Nissen (1931) also describes male chimpanzees carrying infants short distances, and van Lawick-Goodall (1967) observed an older male sibling protecting a motherless infant, and also reports adult males as tolerating and protecting infants (van Lawick-Goodall, 1968).

E. RODENTIA

In the rodents most of the information relates to the myomorphs.

1. Sciuromorphs

For the Sciuromorphs C. C. Smith (1968) found that in the wild *Tamiasciurus douglasii* and *T. husdonicus* females with young never had any association with an adult male. Sollberger (1940) found that males of the flying squirrel *(Glaucomys volans)* did not respond to the cries of the young either in the wild or in captivity, whereas the females did. Alcorn (1940) notes for the ground squirrel *(Citellus townsendii)* that the males appeared to take no part in caring for the young.

The female woodchuck *(Marmota monax)* is said to drive the male from the nest as parturition approaches (Schoonmaker, 1938). However, some males of this group may be tolerant of the young, e.g., the beaver *(Castor canadensis)* (Tevis, 1950).

2. Hystricomorphs

There is practically no information for the hystricomorphs, but Kleiman (1969a) found that in captive groups males of the green acouchi *(Myoprocta pratti)* showed patterns of maternal behavior similar to those shown by the mother, such as grooming and licking the anogenital region, and were very attracted to the young, following them round the cage. The males had limited contact for the first week or so after parturition, as the mothers were very aggressive toward them, but they were very persistent and slept in the nest with the young when the mother was not there. Some males seemed to show an increased tendency to carry nest material to the nest and to dig in it after a recent birth. From Day 2 young were commonly observed to go and sleep in the male's nest, where their attempts to suck from him would typically be responded to by licking. Aggression by males was seen only twice—once when a strange young got in with a male, and once a male killed a young animal in his own group. King (1956) studied guinea pigs *(Cavia porcellus)* under semi-natural conditions and found that although the males offered no care to the young, they were very tolerant of them. Kunkel and Kunkel (1964) studied caged groups of *Cavia aperea* f-*porcellus* and found that males were indifferent to young born in their own groups but appeared nervous of strange young.

3. Myomorphs

The myomorphs include the species that are most commonly used as laboratory animals, and there is some information concerning a relatively small number of species.

von Koenig (1960) found that in captive *Glis glis* the males were excluded from the nest until the young were about 16 days old, but that after that they eagerly entered and maintained the nest and inspected the young, who would sometimes follow the male to a new nest box and stay with him until hunger made them go back to the mother. However, Koenig thought that in the wild the male probably had nothing to do with caring for the young. Nevo and Amir (1964) never found wild *Dryomys nitedula* in the nest with the young, and thought that the female rears the litter unaided.

In the genus *Peromyscus* the males show considerable care for the young, especially for older pups. Possibly they would also show care for younger pups if they were allowed access to them by the mother.

In *Peromyscus maniculatus* a male may or may not be driven off by the female, and this may at least partly depend on whether he is her mate or a strange male. Eisenberg (1963) notes that the males of *P. maniculatus* may remain with the females during parturition, although Eisenberg (1962) found that the males might be excluded for the first day or two. King (1958) found that both *P. m. gracilis* and *bairdii* females with litters attacked introduced males, but tolerated males living with them before parturition. Howard (1949) and Blair (1958), using nestboxes to study a wild population of *P. m. bairdii*, found that the male continued to stay in the nest with the female and her litter, and might associate with the litter outside the nest. In a captive population of the same subspecies, Horner (1947) found that males in the same nest with the female and her young at 26 hours postpartum, and noted that the males nest-built and licked the young. The same author found that *P. m. gracilis* males would carry the young in the mouth and would help the female move the young when the nest was disturbed. One male of this subspecies was seen to pull a fetus from the female's perineum and eat it. A male *P. m. nebrascensis* was seen washing young 24 days old. F. H. Clark (1937), speaking of *P. m. artemisiae*, said merely that a male in the same cage with a female and her litter did not molest the young.

For *P. californicus* Eisenberg (1963) said that in captivity females allowed males to remain with them during parturition, although Eisenberg (1962) described the female as partitioning the nest and excluding the male from immediate contact for 1–2 days postpartum. Eisenberg

(1962) noted that males of this species have never been seen to carry the young. *Peromyscus crinitus* females in captivity drive males away before and after parturition (Eisenberg, 1963; Egoscue, 1964), but the family shares a common nest once the young are 16–21 days old (Egoscue, 1964).

Nicholson (1941) studied *P. leucopus noveboracensis* in the wild, and found it unusual for adults other than the mother to be in the nest with the young, and in captivity he found that the females force the males out at least until the young are 18 days old. Horner (1947) caged a wild-caught male with a female who bore a single pup (by a different male). At 22 days old the pup was removed for 10 minutes, and on its return it was licked by the male as well as the female, and the male carried it back to the nest. Svihla (1932) found that introduced *P. leucopus* males were usually attacked by a female with a litter.

Goodpaster and Hoffmeister (1954) found that a caged male *P. nuttalli* no longer went into the nest once the young were born.

Bailey (1924) noted that in captive groups a male *Microtus pennsylvanicus* may be attacked if he comes near newborn young, and that he may eat young left unguarded. However, the same author implied that a single pair living in a cage may share the nest with the litter. Frank (1952) found that with *Microtus arvalis* in captivity the males often retrieve the young. Yardeni-Yaron (1952) found that captive *Microtus guentheri* males shared the care of the litter with the female and that a high proportion would adopt strange young, and even 50% of naive males would do so. Birkenholz and Wirtz (1965) found that in some cases captive male *Nyctomys sumichastri,* the vesper rat, stayed in the nest with the mother and litter, but usually remained outside until the litter was 1–7 days old. Dieterlin (1960) found in captivity that a male harvest mouse *(Micromys minutus)* was allowed to remain in the nest before and after parturition.

Blair (1941) found that in a captive pair of *Baiomys taylori* the male parent seemed as solicitous of the young as did the female, including carrying back to the nest, licking them and standing over them in the female's absence. de Kock (1966) stated that although most males are attacked by the female before parturition, some male lemmings *(Lemmus lemmus)* will participate in rearing young, but he gives no details. Layne (1959) found that in captivity a male *Reithrodontomys humulis* (eastern harvest mouse) would join his mate in defending the young; Kaye (1961), however, also working with laboratory-living animals, found that the care of the young was left entirely to the female, who drove the male from the nest a few days before parturition and would not readmit him until about 3 weeks after the birth.

Rowell (1961a) tested male golden hamsters (*Mesocricetus auratus*) with pups 3–11 days old which were placed on their cage floors, or in their nests. She found that three out of eight males killed and ate the pups, while five of them retrieved the pups to the nest and licked them, but did not brood them. However, Dieterlin (1959) reported that males living with lactating females (a different situation from that of Rowell) showed "an inhibition of biting" toward young less than 15 days old, and described one male as retrieving and brooding his mate's 1-day-old litter.

Male African giant rats *(Cricetomys gambianus)* were found to remove young from the nest, using behavior comparable with the retrieving of an inexperienced female (Ewer, 1967). Apparently this was, on at least one occasion, when he was trying to copulate with the female.

Beniest-Noirot (1958), working with laboratory mice, reported no difference between males and virgin, pregnant, and lactating females in retrieving, nest-building, nursing or licking, when they were presented with newborn foster pups; the data, however, were expressed as presence/absence behavior, not on its frequency or any measure of intensity. Brown (1953), studying laboratory groups of *Mus musculus* which were descendants of wild animals bred in a stock colony, found that males were sometimes attacked by females before parturition, and, although the males were occasionally observed carrying young (it is not clear why), they were generally not interested in them.

Causey and Waters (1936) tested male laboratory rats with pups and found that although they would approach and smell them, they would not retrieve them. However, they have been found to retrieve, crouch, and lick in response to pups provided a long enough exposure period had been given (Rosenblatt, 1967).

F. LAGOMORPHA

For the lagomorphs there seems to be practically no information on the response of the males to the young, although Grange (1932) noted that in captivity males of the snowshoe rabbit *(Lepus americanus)* seemed to pay little attention to the young.

G. CETACEA

There is only a little information, for one species in captivity. A. F. McBride and Hebb (1948) noted a newborn infant *Tursiops truncatus* being bitten by an aggressive male, but Essapian (1963), also observing captive animals, found that although a male seemed to menace a mother with a neonate there was an abrupt change when the infant was 2 weeks old. The bull was seen "baby-sitting" while the mother fed at the platform—a type of behavior also described for females (see Section III,

9). When another female gave birth the bull accompanied her and the calf, protectively shielding the calf from other animals, a role also often played by a female (see Section III, G).

H. CARNIVORA

1. Pinnipedia

Males of carnivore Pinnipedia generally ignore the young. Thus the male elephant seal (Mirounga angustirostris) is liable to crush the young by crawling over them (Bartholomew, 1952). Laws (1956) noted that a male M. leonina from which a pup tried to suckle was tolerant and moved away. The gray seal (Halichoerus grypus) is reported to crush young by crawling over them (Lockley, 1966), and so is the male Californian sea lion (Zalophus californicus) (Bonnot, 1928). However, Eibl-Eibesfeldt (1955) reported that the Galagagos sea lion male Z. Wollebacki (= californicus) jostles babies shoreward if they enter deep water.

2. Fissipedia

Among the Fissipedia most of the information concerns the Canidae.

In the coyote (Canis latrans) males are known to bring food to their pregnant mates (Cahalane, 1947) and to help the female provide food for the young, although living in a separate burrow (Bourlière, 1955). However, Dobie (1950) said that both parents are found in dens with pups, and that the male shares in the guard duty, being more ready than the female to challenge an enemy and to try to lead it away. This author also said that the male as well as the female will move pups if danger threatens. Possibly there are subspecific differences in behavior as Grinnell, Dixon, and Longdale (1937) reported that C. latrans lestes males bring food no nearer than 100 yd of the den, whereas C. l. ochropus males bring food to the entrance. In the latter subspecies, the males help teach the young to catch prey.

In domestic dogs and wolves, Scott (1950, 1962) noted that both sexes may provide food for the young, and Murie (1944) and Ginsberg (1965) described regurgitation of food by male wolves. Young and Goldman (1944) noted that the wolf (Canis lupus) male may assist the female in digging a den for the young, that he brings food before and after whelping, that he acts as a sentinel, running off if there is an alarm, giving a succession of howls and letting himself be seen. Furthermore, males have been reported as moving the cubs from the den, and also tending them if the mother is killed. Ginsberg (1965) observed that in a captive group a dominant male which had access to the pups helped to carry and clean them. In contrast Yadav (1968) noted that in zoo-living Indian wolves, *C. lupus pallipes,* the male was indifferent to the pups and took no part in rearing them.

Maberley (1966) reports that in the silver-backed jackal, *C. mesomelas* both parents forage for the young, and regurgitate food for them when they are very young. The golden jackal *(C. aureus)* has territories occupied by a male and one or two females, and the male regurgitates food to young (van Lawick, 1969).

Kühme (1965a,b) noted that male African hunting dogs *(Lycaon pictus)* shared in caring for the young, guarding them, calling them, and regurgitating food for them while other pack members hunted. Estes and Goddard (1967) saw 5 adult males feeding an orphaned litter by regurgitation until the pups were old enough to run with the pack. Males also play with young (van Lawick, 1969).

The male European red fox *(Vulpes vulpus)* is primarily responsible for cleaning the den and gets food for the nursing vixen (Westell, 1910; Seton, 1929) and the litter (Sheldon, 1949). The latter also noted that the male plays with the cubs. Other species of *Vulpes* seem to be similar. Thus *V. velox* and *V. regalis* males are both thought to care for the young (Seton, 1910) although no real evidence is presented. Egoscue (1956) noted that the females of the kit fox *(V. macrotis)* spend most of the time in the den when suckling the young, and thought that the males must do most of the hunting during this period. A zoo-living *V. corsac* pair reared young, the male playing an active part, including actively defending the young (Dathe, 1966). A gray fox male *(Urocyon cineroargentus)* was seen with a female and her litter, apparently teaching the litter to hunt (Grinnell *et al.,* 1937).

In otters the evidence seems to suggest that the males play a less active role than in the Canidae. Thus for the sea otter *(Enhydra lutris)* Fisher (1940) noted that although there may be a few males in the nursery areas, they are not interested in the cubs. Crandall (1964) observed that when a *Lutra canadensis vaga* female gave birth in captivity she ejected the male from the shelter. Although he made several attempts to reenter, it was 6 months before he was fully accepted into the family group. A further note (Anonymous, 1956) shows that, again with captive *L. c. vaga,* the male showed immediate interest when the young first emerged, but was chased away by the female. However, Liers (1951a) said that although in the wild the male *L. canadensis* keep well away from the cubs, in captivity they are usually kind to the young, and that the male sometimes adopted orphaned cubs and would catch food for them. Liers (1951b) noted that when the young emerge from the nest the male often joins the group, plays with the babies, and helps to teach them to swim. Grinnell *et al.* (1937) found in California that a male cared for a brood of half-grown young whose mother had been trapped. For the European otter *(Lutra lutra)* the evidence seems to point to a casual attitude in the male (Lloyd, 1951), although Stevens (1957) states that even in the odd instances when the whole family is seen hunting and playing together the male is living in a separate holt, and generally watching the family from a distance.

Male bears *Ursus piscator* are reported to kill the young if they meet them while they are still small (Bergman, 1936).

For the wolverine *(Gulo luscus),* Seton (1910) quoted a note to the effect that the male may help in caring for the young. Male skunks *(Mephitis mephitis)* are reported to kill them (Stegeman, 1937).

Mustela pennanti is said to take no part in rearing the young (Seton, 1910), whereas *M. noveboracensis* is said to be found in pairs when caring for the young, the male having been seen carrying food to a den containing young (Hamilton, 1933). *M. cicognanii* parents are said to be both found at the home den (Bishop, 1923). In the stoat *(M. erminea)* the sexes usually live separately, although there are records of a male associating with the family (Southern, 1964), and in *M. nivalis* the sexes remain separate (Polderbeer *et al.,* 1941). Mink farmers are advised to separate the male from the female and young *(M. vison)* because the male tries to brood the young (Seton, 1910).

The pine martin *(Martes martes)* male probably takes little part in rearing the young (Southern, 1964).

A *Suricata suricata* male in captivity spent much time grooming a litter by fleaing them, although not by licking them in the way their mother did (Ewer, 1963). A note on the binturong *(Arctitis binturong)* in captivity states that the male stayed with the female after the litter was born without any serious disputes occurring (Gensch, 1962). *Helogale undulata* males in captivity (von Zannier, 1965) have been seen covering the young when the mother left them for a short time very soon after the birth. He said that all members of the family, including older male and female siblings, actively participated in caring for the young.

Among the Felidae the lion *(Panthera leo),* males in captivity, are reported as showing no interest when in adjoining cages to newborn litters (Cooper, 1942) but Schenkel (1966) said that, in the wild, males occasionally "mother" cubs. For the bobcat *(Lynx rufus),* Gashwiler, Robinette, and Morris (1961) found no evidence in hunter's records that the male helps with the care of the young, but Young (1958) stated that both parents bring food to the kittens before and after the den is abandoned.

However, the mountain lion male *(Felis concolor)* is thought to have nothing to do with rearing the young (Grinnell *et al.,* 1937). The wild cat, *F. sylvestris,* female rears her young away from the male, which is known to kill kittens (Southern, 1964).

I. HYRACOIDEA

The wild hyrax *(Procavia capensis)* male is excluded from the groups of females with their young (Hahn, 1934), and it was observed in captivity that some males killed the young although others did not (Mendelssohn, 1965).

J. PROBOSCIDEA

In the African elephant *(Loxodonta africana)* living in the wild, a young bull has been observed helping the newborn to its feet, behavior also shown by females and other calves (Poppleton, 1957); and in captivity the Indian elephant male *(Elephas maximus)* has been observed in some cases to be tolerant (Maberry, 1964) and in some aggressive (Anglis, 1964).

K. UNGULATES

1. Perissodactyla

In the Equidae it seems that the male is often aggressive to the foal. Wackernagel (1965) noted that although in zoo-living Grant's zebra *(Equus burchelli)* the stallion sometimes defended the foal, he sometimes attacked it. Klingel and Klingel (1966) observed *E. quagga* and noted that the stallion watches the birth with interest, but is chased away by the mare after the birth. A zoo-living stallion of *E. hemionus,* the Indian wild ass, was reported by David (1966) as rushing at a newborn foal and picking it up by the neck; Tyler (1969) found that stallions in the New Forest generally ignored foals, but were more tolerant of their approaches than mares. Occasional foal-killing by stallions occurred when they had been kept away from mares for the nonmating period and then were released among them. Yearling colts as well as fillies were interested in foals to the extent of standing staring at them.

2. Artiodactyla

In this group the cows often seek isolation, at least from the main herd, and sometimes from all the species members, when about to give birth (see Section III, K), and thus the males may have little opportunity to interact with them. Despite the fact that this order contains a number of species which have been extensively studied in the wild, and several important domesticated species, there is very little information on male behavior toward young.

Maberley (1966) quoted an observation that the giant forest hog *(Hylochoerus meinerzhageni)* male helps the female build a nest when she is about to farrow, piles nest material up over her, and guards the spot. Hafez, Sumption, and Jakway (1962) say that nothing is known of the male's response to young in domestic pigs.

Neal (1959) writing of the collared peccary *(Pecari rajacu)* in Arizona, says that rearing is left to the sow, that the males take no interest, and that the sows with their young are frequently seen feeding by themselves.

Among the antelopes there is also little evidence of male care. Buechner (1950) notes that in the pronghorn antelope *(Antilocarpa americana)* the bucks are mostly away in bachelor herds while the fawns are being born. However he says that a semitame buck who came upon fawns some 18 hours after birth sniffed and pawed at them. Dittrich (1968) describing zoo-living antelope reports that among the gazelles the males of *Gazella soemmerringii* and *G. leptoceros* attack calves, although in other species the males sometimes stand beside a "hidden" calf.

Espmark (1964a) found that reindeer *(Rangifer tarandus)* cows with calves usually keep away from the bull, who attacks calves that come too near.

Male chevrotains *(Tragulus javanicus)* are often seen resting near the fawns, which are left alone for much of the time for the first few weeks (Davis, 1965).

In captivity, male klipspringer antelope *(Oreotragus oreotragus)* have been known to become aggressive toward anyone approaching the hiding place of the fawns (Cuneo, 1965).

L. BEHAVIOR OF MALES TOWARD YOUNG: SUMMARY

In summarizing the information in this section there are four main points to be discussed. First, how widespread is the phenomenon of male behavior to young. Second, how far the male shows the same patterns as the mother, and/or other females in the group. Third, how far the males show behavior peculiar to their sex that contributes to rearing the young. Finally, the sort of factors that affect whether or not, given the opportunity, males will care for their young.

It is difficult to assess how widespread the care of young by males really is, since for such a high proportion of mammalian species little or nothing is known of the breeding biology. However, the survey above indicates that care by the males is by no means uncommon in three of the largest and best known orders, namely the rodents, the fissipede carnivores, and the primates. In a large proportion of the remaining orders there are records of what appear to have been incidents of care for the young: Insectivora, Chiroptera (a doubtful case), Cetacea, pinnipede Carnivora, Proboscidea, Artiodactyla. This leaves only the Dermoptera, Pholidota, Lagomorphs, Tubulidentata, Hyracoidea, Sirenia, and Perissodactyla — all rather small and, with the exception of the lagomorphs and perissodactyls, little known groups.

Clearly when males attack young (e.g., *Tursiops truncatus, Equus burchelli, E. hemionus, Procavia capensis, Mephitis mephitis, Ursus piscator, Rangifer tarandus, Presbytis entellus*), they show a type of behavior not normally shown by the mother. However, the first four of the above examples were based on observations of animals in captivity; under these conditions even mothers may attack their own young. There are many examples of females excluding the males from the maternal nest, e.g., *Peromyscus leucopus noveboracensis,* or bearing their young in isolation from males (e.g., many Chiroptera and ungulates), and thus in many species the male does not have the opportunity of showing any response to the young. However, it should not be assumed that where the male does not have access to young he would in all cases kill them if he did [e.g., see above Buechner (1950) with respect to the pronghorn antelope, and Liers (1951a,b) for the otter *Lutra canadensis*]. His separation from the mother and young may be adaptive for other reasons, such as decreasing the disturbance in and around the nest, thereby decreasing the likelihood of predators being attracted to it. In the case of young who follow their mother from an early age, as in ungulates, isolation may allow the young to learn her specific characteristics, although in view of the numerous opportunities for interactions with other species members which occur in many of these species, this seems unlikely (see Section V).

In many cases the males do care for the young, and the patterns are often very similar to those shown by the female, although obviously the males do not suckle the young. Even if male guinea pigs are induced to lactate by surgical techniques combined with hormonal treatment (Nelson and Smelser, 1933), they do not allow the young to suckle unless held forcibly. Nevertheless, males help dig the nest (e.g., wolves), groom the young (e.g., some rodents), retrieve them (e.g., some rodents), carry them (e.g., *P. hampadryas,* some Fissipedia), brood them *(Mustela vison),* and teach them to hunt (e.g., gray fox). In many cases the males behave similarly to females other than the mother, either in tolerating the young, watching, touching, or cuddling them (e.g., many primates), and it should be noted that the females concerned in this comparison are often lactating ones.

There are, however, species in which the male, shows behavior not shown by the mother in that species, although the role played is played by the mother in other species. Thus in marmosets and tamarins, e.g. *Callithrix jacchus*, it is the male parent who cleans the newborn young and carries it most of the time; the male *Tupaia* at least sometimes is the parent who builds the nest for the young. Division of labor in rearing the young is also shown among those carnivores in which the male does all the hunting (e.g., the kit fox).

The work concerning the effects of hormones on male parental behavior has been reviewed by Lehrman (1961), and there is also some discussion in Richards (1967). The experimental data on the effects of hormones concerns rodents, and includes a very limited number of species. While there is a little evidence for hormones affecting male parental behavior in mammals (e.g., Riddle *et al.*, 1942; Leblond, 1938), there is also evidence for experiential effects, such as caging with young (e.g., Leblond, 1940; Rosenblatt, 1967). Clearly a great deal of further work is needed, and possibly a fruitful line of study might be the factors affecting male parental behavior in species in which under natural conditions the males show care for the young similar to that shown by the female, and in species in which, again under natural conditions, males show care which the female does not.

With the above facts in mind it is not easy to draw conclusions as to what factors might be controlling whether or not males show care for their young. It is clear, however, that each species needs to be investigated individually.

III. BEHAVIOR OF FEMALES TOWARD YOUNG

In some species the young may be reared communally by a group of lactating females (e.g., some bats, carnivores, and rodents). However, in most species a stable and specific mother-infant relationship is maintained from birth to weaning and perhaps longer. This is often done in the face of interactions between the young and other group members which may be initiated either by the young or by the group members (e.g., some ungulates and primates).

There are thus two aspects to the problem of the behavior of females toward young. First, how specific is a mother to her own young, and second, how far do females, lactating or not, show interest in, or care for, young not their own? The questions one should ask about a particular species, or taxonomic group, thus apply in one of the two following groups of situations.

1. A female, with or without young of her own, is approached by the young of another, and may then show care for it, such as allowing it to suckle, or not. This implies initiative being taken by the young who are therefore being nonspecific, but a similar situation arises when an orphaned young is noticed by a female, who may then care for it, or not.

2. The second type of situation is that where young are being cared for by their own mothers, and are in more or less continuous contact with them; and other females, again, with or without their young, may or may not show interest in them or care for them. This implies that the initiative is here being taken by the female, not by the young. As such attentions are often resented by the mother, who will protect her infant from them and therefore prevent the full expression of the behavior involved, they are often difficult to interpret, and it may be wrong to say that such females are behaving maternally. Before doing so one needs assurance that, in the absence of the mother, the females would in fact be maternal. Furthermore, the care shown by females may nevertheless be different from that by the mother.

In neither case, of course, does the behavior necessarily point to nonrecognition of own young by mothers or own mothers by young. Furthermore a female who is specific to her own young while it is present, may not necessarily remain so if she loses it, and the same may be true of the young in their responsiveness to females. Finally, young may not be attractive to females other than their own mothers unless they have lost their mothers and are giving distress calls.

Within a particular species whether the interactions of young with other species members falls into group 1 or 2 will depend partly on the nature of the social organization of the species and partly on the behavior of the young. The division made here is no more than a convenient, but not entirely satisfactory, way of organizing the material.

The data are arranged according to orders; for some orders there seems to be no information.

A. Marsupials

There has been work on the cross-fostering of young in this group. Thus Sharman and Calaby (1964) and Sharman (1962) successfully cross-fostered newborn young of the red kangaroo *Megaleia rufa* and the brush-tailed possum *Trichosurus vulpecula*, respectively, even onto non-lactating virgin or parous animals, in both of which lactation was induced in both species. Clark (1968) found that with *Megaleia rufa* the young generally died unless they were fostered onto mothers whose own young had been less than 20 days different in age. Merchant and Sharman (1966) cross-fostered the pouch young of a number of species of kangaroos and wallabies *(Megaleia rufa, Macropus giganteus, Protemnoden rufogrisea, P. eugenii, Petrogale xanthopus),* foster-young of both the same and different species being accepted by the mothers. However, the interspecies exchanges almost always resulted in the eventual death of the young. Spontaneous fostering of a large pouch young in addition to their own sometimes occurred in the quokka *(Setonix brachyuris)* and in the tammar *Protemnodon eugenii.* Red kangaroos, *Megaleia rufa,* would sometimes inspect the young of other species, but ran away if the young attempted to enter their pouches. However, this species did occasionally accept a suckling post-pouch young of the same species. Hediger (1958) observed acceptance of an additional large pouch young in zoo-living Bennet's wallabies.

Bandicoots *(Perameles nasuta)* living in small groups in fairly large enclosures were observed by Stodart (1966), who found that the young generally followed only the mother, and all other animals avoided them. However, in two instances when two litters were born a few days apart their weights indicated that they might have changed pouches, and later they were seen to follow either of the two females.

B. INSECTIVORA

Probably the young are almost always reared by the mother alone in this order, and thus no opportunity is afforded for behavior toward them by females other than the mother. However, this does not necessarily mean that the females would not interact with young, given the opportunity.

Conoway (1958) working with laboratory-living least shrews *(Cryptotis parva)* found that adult females other than the mother cared for the young, but he did not specify any details of the behavior and remarked that one litter died from starvation because of disturbance in the nest. However, when a litter was deliberately scattered, all adults participated in rebuilding the nest, and all carried the young.

C. CHIROPTERA

In these group-living animals, the situation with respect to the relationship between mothers and young falls into category 1. The young are generally born in nursery colonies (e.g., Deansley and Warwick, 1939), where they are left while lactating females forage, although some bats carry their young, at least for the first few days (e.g., Mohr, 1933). On her return, each mother must either seek out her own baby among the clusters of young, or nurse any young indiscriminately. The records suggest that behavior in this respect varies with the species, although Brosset (1966) stated that it is certain that in the great majority of species mothers rear their own young.

Mothers seem to be unspecific in the Mexican free-tailed bat *(Tadarida brasiliensis)*, according to Davis, Herreid, and Short (1962), who watched females returning to the nursery area. They thought that mothers accept and are suckled by the first two babies to find the mammae. Furthermore they tested mothers with a series of randomly selected babies, and found that any baby was accepted. Brosset (1966) recorded similar nonspecificity for *Miniopterus schreibersi*. Gates (1941) found that in a captive colony of *Nycticeius humeralis* the mothers were not specific as to which young they nursed, once the young were no longer carried by the mother on her foraging flights. Roth (1957) was uncertain whether in *Myotis lucifugus* the mothers recognize their own young, and found that when young were switched from female to female in cages, they remained in good health, suggesting nonspecificity on the part of the mother. However, this does not necessarily mean that mothers would not be able to recognize their own young, nor that they reject strange young in favor of their own if given a choice. Eisentraut (1936) thought that *Myotis myotis* recognize their own young, while Castaret (1938, 1939) thought that females nurse any young. Kleiman (1969b) found that in captive pipistrelles *(Pipistrellus pipistrellus)* a few females had alien young attached, but Racey (1969) reported that, in cages containing large numbers of mothers and babies, young which have been rejected by their mothers are not accepted. In this case, there may have been a defect in either the rejected young or their mothers, and the evidence does not altogether exclude the possibility that a mother whose young had been removed would adopt another.

Coming now to the species in which the mothers do seem to be specific to their own young, Beck and Rudd (1960) in a brief note said that in a nursery roost of *Anthrozus pallidus* each female seemed to handle her own particular young. O. P. Pearson *et al.* (1952) found that lactating females of *Corynorhinus rafinesqui* climbed among the juveniles apparently searching for their own. That they did find them was confirmed by banding, and acceptance of their own young and rejection of others was found when young were dislodged, and then investigated by the females. Davis, Barbour, and Hassell (1968) described lactating *Eptesicus fuscus* as apparently searching for their own young, and confirmed by banding that they did in fact find them. Observations of captive animals have suggested specificity in the mothers of a number of other species. Thus Wimsatt and Guerriere (1961) found that in the vampire bat *(Desmodus rotundus)* the more mobile young which were no longer attached to their mothers for most of the time were sometimes bitten and killed by the older animals, and these attacks were thought to be provoked by attempts to nurse the wrong adults. In the noctule *(Nyctalus noctula)* Kleiman (1968a) found that females would attack alien young, whereas in the serotine *(Eptesicus serotinus)* such young were not attacked, although females seemed able to distinguish their own young and promiscuous suckling was never seen.

D. PRIMATES

In this group the young of a few prosimian species are deposited in nests, but in all the Anthropoidea the young are carried from birth by the mother or, in the marmosets and tamarins, by the male. Mother-infant pairs interact closely (see DeVore, 1965), and the situation with respect to other group members falls into category 2 discussed above, although there is some information on situations in which females were exposed to motherless infants.

The general response to young by group companions is intense interest, in the form of looking at and trying to touch the newborn, particularly by the females, from which the young are often protected assiduously by their mothers. This makes the behavior shown by group companions difficult to interpret, but evidence from the species in which the mother is permissive, from observations on orphans, and from experiments indicates that such animals are attempting to cuddle and carry the babies, although in rhesus monkeys anyway their behavior may not be entirely like that of the mother (see below).

1. *Prosimii*

Kaufmann (1965) twice observed that two females of tree shrews, *(Tupaia glis)* in captivity placed their young in a common nest, but one or both litters died and no information is given about specificity of suckling.

In a field study, Petter (1965) found that *Lemur macaco* females were allowed by the mothers to lick their infants, and he also notes that in the Indriidae newborn young were of great interest to group members, who tried to lick them.

Doyle *et al.* (1967) described *Galago senegalensis* living under semi-natural conditions. A female other than the mother showed only brief interest in and grooming of the baby at one birth, and no interest was seen in a pair of twins born in a group. However, Sauer (1967) observed another captive group of this species (but a different subspecies) and found that one particular female often took the mother's place in carrying the infant, and also shared the nest with the mother and infant and licked and groomed the latter. This female also retrieved the infant to the nest when he signaled distress, but unlike the mother did not do so at other times. Gucwinskand and Gucwinski (1968) also observed intense interest in a young zoo-living *G. senegalensis zanzibaricus*, the group members rescuing it off the wire as it climbed. Sauer (1967) described how, when a *G. crassicaudatus* female with a young baby was separated from her previous companions, the females (and males) were intensely interested in the baby and pressed against the wire adjacent to the nest box. Crandall (1964) reported that in captive *Nycticebus coucang* the young were as liable to be found clinging to an older sibling (sex unspecified) as to either of the parents.

2. Anthropoidea

Among the Callitrichidae the males are often responsible for much of the care of the young, but other animals may also give care, at least in captivity. Thus Hampton *et al.* (1966) noted much interest by a juvenile female in twins born into a group of *Oedipomidas oedipus*. However, she would attempt to dislodge them if they got on her back. As the male normally carries the young in this species anyway, the interest shown by the female is difficult to interpret. Certainly her rejection of the infant from her back should not be regarded as evidence that she was not behaving "maternally" when she showed interest in it.

In captive *Callithrix* (= *Hapale*) *jacchus*, although it is again the male which normally carries the young most of the time from birth, older siblings (of unspecified sex) were also reported to do so by Lucas, Hume, and Henderson Smith (1937), although not in the first 3–4 weeks of life.

Epple (1967) reported that in captive colonies of *Saguinus geoffroyi* the infant may be carried by members of the social group other than the parents.

Bernstein (1964) made general comments on monkey colonies of various species, and said that among Cebidae (species unspecified) a mother will nurse an infant other than her own, the infant largely determining with which adults it will be associated; however, others who have worked with this family imply that much of the initiative is taken by the females rather than the infant.

In the squirrel monkey *(Saimiri sciureus)*, at any rate in captivity, not only do female group companions show considerable interest in neonates, sometimes trying to drag them off the mother (Bowden *et al.,* 1967), but relatively long-lasting relationships may be established between an infant and a particular female (Hopf, 1967). Such a female continually seeks proximity and contact with the infant, and her behavior after the first 3 weeks of the infant's life can be distinguished from the general attention given by other adults, and she remains tolerant of it after other adults have ceased to be so.

In the howlers *(Alouatta palliata)* on Barro Colorado, a female may be the focus of attention of several other females during parturition (Carpenter, 1934, 1965). After birth the newborn attracted other females (Altmann, 1959), particularly those without young of their own (Carpenter, 1965), and they touched it with their muzzles and hands, despite mild avoidance by the mother. Carpenter never observed females to carry infants away from the mothers, although they might pull at them when they were off their mothers.

Similarly Williams (1967) reported that in a captive colony of woolly monkeys *(Lagothrix lagotricha)* all other group members crowded round a mother during parturition. Five to six days after birth the mother allowed them to touch the infant.

Struhsaker (1967a) observed free-ranging *Cercopithecus aethiops* and found that the infants, especially the newborn, were handled, groomed and carried extensively by females aged 6 months to 4 years, and less frequently by adult females. The observation that an infant less than 14 hours old was carried for nine consecutive hours for distances up to 30 feet from the mother is reminiscent of the behavior of langurs. In caged *C. aethiops* and *C. nitis* Booth (1962) described adult females as picking up any infant that was left alone, although when the mother was present she did not allow them to do so. Bernstein (1964) stated that in a captive group infants were often seen being held by an older female sibling. When a patas *(Erythrocebus patas)* infant was born into a small captive group, an adult female showed intense interest in it and was allowed by the mother to carry it about after the fourteenth day (Hall and Mayer, 1967).

In wild chacma and olive baboons *(Papio ursinus* and *P. anubis)* Hall and DeVore (1965) described the adult females as not showing interest in newborn infants, but Bolwig (1958) and DeVore (1963) described intense interest by the whole troop as soon as a birth was noticed. The females tried to touch the infant, and later were allowed to grasp it (DeVore, 1963). However their interest waned rapidly when the infant was 4–6 months old, unlike that of the males (see Section II, D). Bolwig (1958) noted that adult females and juveniles were interested in young and that the juveniles might carry them. Ransom and Ransom (1969) reported that adolescent females and adult females without babies of their own will sometimes carry infants. Gillman (1939) made some observations on chacma baboons living under restricted conditions in captivity and noted adoption of infants by mature females. However, the females molested foster infants who tried to feed near them.

For hamadryas baboons (*P. hamadryas*) interest of females in babies seems much less intense than in the other species. Thus Kummer (1968) described females as sniffing at young infants, but apparently they did not attempt to carry them and did not pick up motherless infants (Kummer, 1967).

Free-ranging rhesus monkeys (*Macaca mulatta*) have been studied in India by Southwick *et al.* (1965) and on Cayo Santiago (Kaufmann, 1966). The former described adult females as showing no interest in newborns until they were 6 days old, when interest became intense. Rowell, Hinde, and Spencer-Booth (1964) and Spencer-Booth (1968), working with captive animals, have also described intense interest by females in babies born into their group. The females watched and touched the babies, but would also cuddle and carry them if permitted to do so. Kaufmann (1966) and Spencer-Booth (1968) found that young females, and particularly those who had not given birth, were the most persistent. However, in this species the mothers generally try to protect their babies from such females (Hinde *et al.*, 1964). Even when females can interact fairly freely with babies in a group situation, their behavior may appear ambivalent (Rowell *et al.*, 1964) and they may not completely replace the mother during her absence (Spencer-Booth and Hinde, 1967). In both these respects, however, their behavior may be being affected by the response of the infant (see Section VII, C). Nevertheless, adoption in this species has been observed under laboratory conditions (Harlow *et al.*, 1963; Hansen, 1966), even to the extent of the production of apparently normal milk by nonpregnant, initially nonlactating females.

Simonds (1965) observed that in wild *M. radiata,* the bonnet macaque, other females approached close to a mother and infant, but were not allowed to hold it. In a zoo-living group a female kidnapped a week-old infant, although she had an infant of her own and later began to neglect the foster infant (Bullerman, 1950).

In the above species of primates most of the care is shown either by the mother, or a male, and the other females of the group, despite showing interest, are not normally permitted by the mother to do more, at least when the infant is very young. That their interest is in fact in showing care is demonstrated by a few observations only.

In langurs *(Presbytis entellus)*, however, it is normal for female group members to be readily allowed to hold an infant within a few hours of birth (Jay, 1962, 1963, 1965; Sugiyama, 1965b, 1967), and similar behavior was observed in a small zoo-living group of *P. obscura* (Badham, 1967a), in which the three females passed the two babies around during the first day of life. Such observations do not indicate that mothers do not recognize their own babies, but merely that they are interested in others than their own.

It was reported in *Zoonooz* (1964) that newborn colobus monkeys *(Colobus colobus)* were handled almost from birth by adult and subadult females in the colony.

Reynolds and Reynolds (1965) reported that wild chimpanzee juveniles were sometimes carried by females other than their mothers, and van Lawick-Goodall (1965, 1967) found that adults and juveniles would touch young babies and that older female siblings sometimes showed intense interest in babies, and might carry them, and show "maternal protective behavior." An orphaned baby was "adopted" by an older female sibling. Schaller (1965) described how wild mountain gorilla *(Gorilla gorilla)* infants sometimes appeared to have close social bonds with females other than their own mothers, and that mothers would protect their newborn infants from the attempts by other females to touch them. Carpenter (1940) found in wild gibbons *(Hylobates lar)* that competition for the baby among the females was limited to young maturing females.

In humans there has been some investigation of the stimulus characters of babies which affect responsiveness to them (Gardner and Wallach, 1965) and the development of the attachment of the infant to its mother (Section IV, B), but not on the specificity of the response of women to children. More precise knowledge of such behavior would be very valuable in the context of the fostering and adoption of children.

E. RODENTIA

It is on a few species of rodents that much of the work on the control of "maternal behavior" has been done. There remains a vast number of rodent species for which little or no information is available. Many live solitarily or in mated pairs, and thus the opportunity for females other than the mother to interact with the young does not arise. However, this does not mean that they will not accept them under laboratory conditions, and the golden hamster *(Mesocricetus auratus)* is an example of a species, thought to be solitary in the wild (Rowell, 1961b), in which care of foster young can be induced in lactating and nonlactating females.

The behavior of females toward young ranges from aggression through tolerance, to active attempts to steal strange young. The cases of aggression are often by nonlactating animals, and on the whole mothers seem to be relatively unspecific in their responsiveness to young.

1. Sciuromorphs

There is little that can be said about this group. Tevis (1950) noted that in wild-living beavers *(Castor canadensis)* yearlings were tolerant of the young snuggling to them. King (1955) studied a colony of prairie dogs *(Cynomys ludovicianus)* and found that the lactating females drive other females out of their nests. However, when the young emerged, adult females would groom them, and might accept them in their burrows. Even above ground, females would allow strange young to suckle.

2. Hystricomorphs

Mountain viscachas *(Lagidium peruanum)* were twice seen by O. P. Pearson (1948) to nurse two young of different sizes simultaneously, and either these must have been successive offspring or one must have been from another female.

Kunkel and Kunkel (1964) studied guinea pigs *(Cavia aperea* f. *porcellus)*, and found that females may either reject or care for strange young. In large groups the young were mothered indiscriminately by all lactating females, but in small groups there was no indiscriminate adoption. King (1956) studied *Cavia porcellus* under seminatural conditions and also thought that some natural fostering probably occurs. It seems that in any case mothers are rather casual about their offspring, as Allen (1904) found it was rare for a mother caged in sight of her newborn litter to attempt to reach them.

In the green acouchi *(Myoprocta pratti)* in captivity Kleiman (1969a) found that nonmaternal females would groom the young and lick the anogenital region. Such behavior was seen in pregnant and nonpregnant adults, including nulliparous females. Adults grooming one another were almost never seen to lick, so the licking seems to have been a response specifically to young. On the one occasion when a pregnant female was given a pup from another group, she attacked it. Cross-fostering was possible, lactating females accepting alien young after a few minutes of investigation, indicating that they noticed the difference between own and foster young.

In captive agouti *(Dasyprocta aguti)* apparently several females sometimes give birth in the same nest (Roth-Kolar, 1957), and the mothers tolerate their own older litters suckling at the same time as their new litters. However the young are bitten away when they approach conspecifics other than their own mothers.

3. *Myomorphs*

Meyer and Meyer (1944) found that the cotton rat *(Sigmodon hispidus)* would rear foster young with their own litter, if they were all about the same age. P. G. Pearson (1952), also in the laboratory, observed a female wood rat *(Neotoma floridana)* nursing her two litters, born 42 days apart, simultaneously: this is not fostering, but it does imply some tolerance of young other than the latest litter. Furthermore one of the older litter was observed cleaning one of the younger litter. Egoscue (1957) worked with a laboratory colony of the desert wood rat *(Neotoma lepida),* and found that females would readily accept foster young up to 2 weeks old.

Peromyscus spp. seem to be unusual among rodents in that not only have females been observed to rear litters communally in the laboratory, but also to do so in the wild, although not in all species studied. Thus Howard (1949), using nest boxes, found three cases out of 185 in which two litters of *P. maniculatus* were being reared simultaneously by two mothers, and thought it probable that they nursed each other's young. In the laboratory Blair and Howard (1944) showed that female *P. m. blandus* often stole each other's young, if in different nests, and nursed them if in the same nest, as did *P. polionotus leucocephalus.* In contrast to their behavior in captivity, Blair (1951) reported *P. p. leucocephalus* females as being intolerant of one another in the wild. King (1963) reported that two successive litters of *P. maniculatus* may compete for the mother's milk. Barbour and Gault (1953) report a find in the wild of two lactating females of *P. m. bairdii,* four young (at two different stages of development) and a pregnant female, all occupying one nest. R. M. Hansen (1957) described a nest of *P. maniculatus* containing two females both suckling young of which one removed all the young to a new nest after disturbance. A female was reported as adopting house mice, larger than her own offspring (Ackert, 1910). *P. nuttalli* in captivity were also found tolerant of young other than their own, by accepting them into their own litters (Goodpaster and Hoffmeister, 1954).

Dieterlin (1962) studied spiny mice *(Acomys cahirinus)* in captivity and found that not only did other females nurse the young, including females which had not yet given birth, and at any phase of their estrous cycle, but that females also "helped" at the birth, and often ate the placenta.

Dieterlin (1959) found that in golden hamsters nonmaternal females varied in their response to pups; they might either tolerate or kill them.

Richards (1966a) found that lactating females accepted foster pups, whereas pregnant ones showed maternal responses only after initial attacks, and virgin females attacked them. Rowell (1961a) presented adult females with pups and in only a few cases were they "mothered." Juveniles showed some "maternal behavior" but this was described as qualitatively different from that of adults. Eibl-Eibesfeldt (1953) found that hamster *(Cricetus cricetus)* females that had no young of their own would retrieve very young pups.

Frank (1952) describes captive female voles *(Microtus arvalis)* as retrieving the young other than their own, particularly postparturitive females. Yardeni-Yaron (1952) studied laboratory-living *M. guentheri* and found that females would accept strange young, but that only a low proportion of naive females would do so.

Rosenblatt and Lehrman (1963) found that shortly after parturition female laboratory rats *(Rattus norvecigus)* will retrieve a wide variety of objects having characteristics differing greatly from the properties of newborn young, and Denenberg *et al.* (1966) successfully fostered litters of laboratory mice on rats whose own litters were removed. Beach and Jaynes (1956a) showed that although lactating females will retrieve alien young, they seemed to prefer to retrieve their own young given a choice, and when pups were offered alone alien young were retrieved more slowly than their own young. Wiesner and Sheard (1933) found that nulliparous rats usually (84%) failed to retrieve pups, but the length of exposure to pups is important (see Section VII, C) as Rosenblatt (1967) induced retrieving, licking, crouching and nest-building in nearly all nulliparous females provided a long enough exposure period was allowed. Moltz *et al.* (1966) found that rats whose own pups had been delivered by cesarean operations and removed shortly before full term would retrieve and rear normally delivered foster pups. Denenberg *et al.* (1966), and Denenberg *et al.* (1962) fostered young with success varying according to such factors as age of test pups and the length of time the female had been lactating, although survival was always less than when the pups were reared by their natural mothers.

Brown (1953) studied house mouse *(Mus musculus)* populations consisting of the descendants of wild animals bred in a stock colony and found that females in the late stages of pregnancy, or recently parturient, excitedly licked, sniffed, and handled the young of other mothers: he also found mild interest by nulliparous females. A female which had lost her own litter joined another female in nursing hers, but these were not reared. Eibl-Eibesfeldt (1950) noted that although house mice may retrieve alien young, they often eat them subsequently.

Beniest-Noirot (1962) was able to foster pups on to lactating females of laboratory mice, although the age of the test pups in relation to the females own litter was important. Virgin females also accepted pups, and showed retrieving, licking, nest-building, and lactation position responses to them (Noirot, 1964a).

F. LAGOMORPHA

"Maternal behavior" even by the mother does not seem specially well developed in this group. Thus Ross *et al.* (1959) found no maternal retrieving of neonates by domestic rabbits, although they do nest-build and show some aggression when disturbed (Denenberg *et al.*, 1958).

G. CETACEA

There is some information for one species in situations 1 and 2 discussed above. The data concern the bottle-nosed dolphin *(Tursiops truncatus)* in captivity. Thus Essapian (1963) reported that other females approached a female in labor, looking at her underside, and were present at the birth. One particular female then shielded the infant on one side while the mother did so on the other side. A. F. McBride and Kritzler (1951) report that the mother of a newborn infant is almost invariably accompanied thus by a mature female, and that the latter may help the mother ward off attacks by adult males. Females are also reported by these authors to assist the mother in raising stillborn young to the surface. Essapian (1963) noted that interest in the baby by other animals might persist for several weeks, although Tavolga and Essapian (1957) found that mothers repeatedly retrieved their young and prevented them from approaching other dolphins. However, they also found that infant groups were sometimes left with one female while other females fed; a female who had fed would then change stations with the female accompanying the young. No female was seen to mistake another infant for her own, on return to the group. That the female left with the young was acting protectively, in this case to young none of whom was her own, was suggested when a pregnant female guarding the young clapped her jaws, in the manner of mothers when their young are interfered with, when a male swam between her and the young. McBride and Kritzler (1951) reported that two pregnant females protected an orphaned young dolphin who approached them.

The situation thus seems to be that interest in and care for young is shown by females other than the mother. The mother, while remaining specific to her own young, may nevertheless leave it with other females while she feeds. Females may behave protectively to infants other than their own, whether or not the mothers are present.

H. CARNIVORA

1. Pinnipedia

In this group the pups remain on the breeding grounds, sometimes in very large groups, while the mothers forage, and the situation thus falls into category 1.

Thus Bartholomew (1952) reported that the Northern elephant seal (*Mirounga angustirostris*) would rebuff strange young, although, unlike the males, females were careful not to crush young. However, Klopfer and Gilbert (1966) said that females were very permissive to alien pups, and would readily adopt others, whether or not their own was at their side. Fogden (1968) reported that females would allow two pups of quite different ages to suckle simultaneously, and a pup might suckle two females within a few minutes. Carrick *et al.* (1962) thought that strange pups were allowed to feed only if they remained silent, the females having difficulty in discriminating between pups, but Klopfer and Gilbert say that they are allowed to feed even if they are vocalizing. However it seems that mothers can recognize their pups as Klopfer and Gilbert (1966) removed pups from their mothers experimentally and found that although females with their own pups present replied to the isolated pup's bawling as often as did its own mother, and might even approach it, it was generally the mother that made the more sustained effort to reach it. This is not unequivocal evidence, however: one must show a difference between the responses of two lactating females, one of whom is the test pup's own mother, or offer a female a choice of two pups, one of which is her own, to be absolutely sure that discrimination is occurring. Laws (1956) reported that elephant seal *(M. leonina)* cows remained in the vicinity of their calves until weaning and would very occasionally adopt another pup if they lost their own. Cows often bit a strange pup that came too close, although a cow was once seen suckling two pups, one of which was assumed to be an alien. Preble (1923) thought that in the Alaskan fur seal *(Callorhinus ursinus)*, each female picked out her own pup to nurse, and Bartholomew and Hoel (1953) found that, of seven mother-infant pairs watched for 7 weeks, in only two instances two pups simultaneously but briefly nursed from the same female. Otherwise cows nursed only their own pups and vigorously repelled suckling attempts by others. For the gray seal *(Halichoerus grypus)*, Lockley (1966) quotes E. A. Smith to the effect that cows may tolerate suckling by one or more alien young, although they sometimes drive them off. Cows are described as working their way through groups of young, smelling, and then pushing aside or ignoring them until they found their own. Fogden (1968) and E. A. Smith (1968) found that some females fed more than one pup, and would feed alien pups in the temporary or permanent absence of their own. E. A. Smith also observed that a cow suckling her own pup might be vigorously dispossessed by another cow, which would take over the feeding temporarily.

Rowley (1929) and Mathison *et al.* (1962) thought that the Steller's sea lion cows *(Eumetopias stelleri-jubata)* tended to return to their own pups. Rowley describes cows as walking into bands of pups, nosing about until they find their own. Pups were liable to be killed by the trampling of cows searching for their own pups, and were thrown into the air when they attempted to nurse the wrong cow. Similarly in the southern sea lion *(Otaria byronia)* cows distinguish their own pups (Hamilton, 1934). The returning females bellowed and apparently distinguished the voice of their own pups among the numerous responses, and vigorously repelled attempts by strange pups to nurse.

The Californian sea lion cows *(Zalophus californicus)* also hunt for their own young on coming ashore, and Bonnot (1928) noted also that they treated alien pups roughly. Although several cows approached a squalling, half-drowned pup, they left it after touching it. Similarly Eibl-Eibesfeldt (1955) reported that in the Galapagos sea lion *(Z. wollebacki = californicus)* mothers responded only to the calls of their own young.

2. Fissipedia

Among the fissipedes, females other than the mother may help care for the young in species in which the young are born in a social group; the behavior thus falls within category 2.

Murie (1944) observed a wolf *(Canis lupus)* den containing pups with five adults, and found that one female which was almost certainly not the mother of any of them was especially attentive to them. Ginsberg (1965) found in a zoo-living group that a female "acted as dry nurse."

In the coyote *(Canis latrans)* lactating females may sometimes adopt young. Dobie (1950) reports that two litters may be found in one den, and in one case pups of different ages were being suckled by one female. A female whose own pups had been killed was recorded as stealing domestic dog pups from a camp. Nonlactating females of this species also care for young, at least in captivity, in the presence of their own mother. Thus Kleiman (1968b) observed a yearling nulliparous female showing many of the patterns shown by the mother, such as licking the perineal region and eating the excreta, playing with the pups and curling herself around them in the nursing posture. A yearling female was also seen regurgitating food to pups (Kleiman, 1970).

In the domestic dog there is evidence for individual variation between bitches in their readiness to adopt pups. Menzel and Menzel (1953) reported that whereas some readily adopt puppies even of other strains, others will accept only their own young.

Kühme (1965a,b) found that in a pack of African hunting dogs *(Lycaon pictus)* consisting of six males and two females, both of whom had pups, the females competed in caring for the pups, they both nursed all the young, and the food supply was treated communally. Maberley (1966) reported that the bitches leave the main pack at breeding time, and that several may give birth to pups in a communal earth.

In the arctic fox *(Alopex lagopus)* in captivity a pregnant female followed and retrieved cubs and lay around them in the nursing posture. When two litters were present, both mothers were seen retrieving each other's cubs (Kleiman, 1968b).

Erickson and Miller (1963) observed in the wild two litters of the brown bear *(Ursus arctos)* which happened to mingle. At first the mothers fought, then both went to the rescue of one cub. Finally one female went off with this, and the other stayed with the remaining cubs for at least 2 weeks. However, it was thought unlikely that adoption is commonplace in this species.

von Zannier (1965) reported that in captive *Helogale undulata* half-grown juveniles of both sexes took part in caring for the young, including covering them in the nest.

There seems to be very little information for the Felidae, and perhaps this is because the females give birth away from other species members. However, Maberley (1966) included a quotation to the effect that lionesses *(Panthera leo)* may sometimes act as nurse guards for each other's cubs, and Schenkel (1966) noted that all adult members of the pride were tolerant toward cubs, but that mothers preferred their own cubs to those of other females. In the domestic cat, Ewer (1959) noted that a mother, meeting another's litter, would lick their perineal regions but would spit and drive them away as soon as she came into contact with their heads. However, when two kittens were exchanged between litters (Ewer, 1961) they were accepted by the mothers. Schneirla *et al.* (1963) carried out experiments in which they removed kittens at parturition and then returned them to the females 7, 12, or 15 days later. If only one such kitten was returned the female would accept it after a delay, but if three were returned at 15 days postpartum, the female attacked them or avoided contacts. A littering female with kittens usually accepted strange kittens not too much older than her own.

I. Hyracoidea

Mendelssohn (1965) and Hahn (1934) found that single females of the Syrian Hyrax *(Procavia capensis)* may sometimes be found accompanied by large numbers of young, but apparently do not show any care for them. It is rather doubtful whether this could be considered as non-specificity on the part of the mothers.

J. Proboscidea

Intense interest in newborn young has been reported for both Indian and African elephants. Thus Maberry (1964) described the birth in a zoo of an Indian elephant *(Elephas maximus)*. The other elephants were separated from the mother during parturition, but bellowed and tried to force their way to the mother as soon as the baby was born. When they were let in the females were described as being very protective to the baby and took turns to have it between their forelegs. After both this birth and another, a particular female seems to have been specially attentive to the baby. Poppleton (1957) describing a wild herd of African elephants *(Loxodonta africana)* noted that the females which had gathered round a newborn calf nudged and pushed the calf apparently to try and get it on its feet. Douglas-Hamilton (1969) saw a female with her own calf allow another calf to remain near and to suckle from her.

K. Ungulates

1. Perissodactyla

Among the Equidae, zebras seem to show no interest in young other than their own (Wackernagel, 1965, for *Equus burchelli;* Klingel and Klingel, 1966, for *E. quagga)*, although the former, in captivity, were re-ported occasionally to allow a second foal to suckle as well as their own.

Tyler (1969) has studied free-ranging New Forest ponies and found that although the females sometimes left the group to foal, or were left behind while foaling, they usually gave birth with ponies all around them, and in any case rejoined the group as soon as the foal could move. Adult mares seldom showed any interest in foals other than their own, although yearling females would approach and stare at them. Mares never allowed foals other than their own to suck them.

2. Artiodactyla

A frequent pattern among artiodactyls seems to be that the mothers withdraw from the social group, or form maternity groups, before parturition. They may remain in isolation for a little time after the young have been born, and in some species the young may be left alone by the mother while she feeds. It seems fairly common in this group that once the young are mobile, one or two females are found with a relatively large group of young whose mothers are elsewhere; although such females are often referred to as "acting as nursemaids," there is generally no evidence that they show any sort of care for the young. It would be very interesting to know more precise details of the interactions within such groups, and, in particular, how they form and disperse.

Stegeman (1938) observed that in the European wild boar *(Sus scrofa)* in the National Forest in Tennessee the young were born in a secluded spot and cared for apart from other hogs until able to travel with the group. In the domestic pig fostering can be accomplished by giving a small litter to a sow who already has a small litter, especially if this is done within 48 hours of parturition (Hafez *et al.,* 1962). The sow's ability to distinguish her own piglets apparently increases with their age. Identification may be by olfaction, and fostering may be aided by smearing the foster piglet's skin with the sow's excreta, or by confining them for a few hours with the sow's own litter.

Hippopotamus cows *(Hippopotamus amphibius)* retire from the herd to give birth, and probably rejoin a group when the calf is a few weeks old (Laws and Clough, 1966).

Koford (1957) found that in the vicuna *(Vicugna vicugna)* during the first hour after birth, which tended to take place away from the band, other females inspected the newborn, craning their necks to put their noses to it. By the time young were a day old, however, adults were hostile when approached by them. Pilters (1954) studied zoo-living tylopods and noted that guanaco and llama females suckled strange newborn young for the first few hours of life. A few days later females always warded off strange young.

The young of the Cervidae, which have been born apart from the herd, are often left hidden by the female for days or weeks.

In the red deer *(Cervus elephas)* each pregnant hind goes off on her own at calving time (Darling, 1937, 1938) and is later able to distinguish her own calf.

M. Altmann (1952, 1956, 1960, 1963) found that in the elk *(Cervus canadensis nelsoni)* the preparturient female goes into isolation, driving off the yearling which may still be accompanying her. Later, in the nursery herds one or two cows may remain with the young while the other females graze. Harper *et al.* (1967) studied free-ranging Roosevelt elk *(Cervus canadensis roosevelti)* and found that cows left the herd a few hours before giving birth and returned when the calf could move about freely at about 2 weeks of age. They described the calves as being left with a "baby sitter," namely one cow, while the rest of the herd fed or rested. They implied, but without details, that this cow keeps the calves together actively. They also mentioned harassment of calves by adults. M. Altmann (1956, 1960, 1963) observed that when a disturbance occurred one elk cow would go noisily toward the intruder, while other cows and yearlings would lead the calves quietly away. As with other ungulates, it would be interesting to know how such groups form. Cows recognize their own calves (Altmann, 1963) and, although any mature cow would rush up to a calf giving the distress call, it lost interest once the mother arrived. M. Altmann (1960) and Struhsaker (1967b) observed yearlings suckling from cows, presumably those which had lost their calves. In the absence of its own calf therefore, the elk mother apparently does not remain specific to it.

Linsdale and Tomich (1953) described the mule deer *(Odocoileus hemionus)* as retiring into hiding to give birth, and subsequently driving all others, including yearlings, away from the fawns. However, W. R. Hansen (1958) observed a yearling and fawn suckling together, and it seems likely that the yearling was an older sibling. Golley (1957), who observed penned mule deer, found that does were antagonistic to fawns other than their own, kicking at a strange fawn after sniffing it. Severinghaus and Cheatum (1956) suspected that adoption of strange fawns sometimes occurs in the white-tailed deer *(O. virginianus)*.

In the moose *(Alces alces)* the cows chase away the yearling before parturition, and the cows and calves are solitary, remaining in close contact with each other (Geist, 1963).

Caribou cows *(Rangifer arcticus)* separate into maternity herds, then bear their calves a little apart from other cows and eventually form nursery bands, in which strange young are chased away by mothers which have their own young (Pruitt, 1960; Lent, 1966). However, females that had lost their fawns attempted to kidnap any fawn (Pruitt, 1960), and adoption of lost fawns by such females was believed to be quite common. Lent (1966) observed that nonbreeding cows most commonly avoided young that approached, and although he described cows as being harried by yearlings it seems likely that these were animals trying to suckle, as such behavior was seen.

Prior (1968) describes an interesting observation by Matula, supported by a series of excellent photographs, of an adult roe deer female *(Capreolus capreolus)* attending another female in labor. She struck at the parturient female, but was apparently "in no way malicious." She then helped lick the fawn clean, and both does ate the afterbirth.

Talbot and Talbot (1963) studied the wildebeest *(Gorgon taurinus)* in East Africa and found that if birth took place in the herd, other females would gather round the cow in labor. However, calves approaching or attempting to nurse the wrong cow were butted away, and Estes (1963) observed this whether or not the cows had calves of their own; although cows were more tolerant of older calves, Estes thought that a lost calf would have little hope of adoption in this species.

Cuneo (1965) observed in a zoo-living group of klipspringer antelope *(Oreotragus oreotragus)* that females who did not have young would go up to a "hidden" fawn and lick it. Dittrich (1968) gave some discussion of zoo-living antelope, particularly gazelles, and says that they generally will not suckle the young of another female of the same species, although a *Gazella soemmerringii* female was once observed to do so. Furthermore, they will sometimes attack a newborn calf. He described this specificity of suckling as being unlike the bush bucks, greater kudus, dik-diks *(Madoqua kirkii),* waterbuck, *(Kobus megaceros),* and sunis *(Nesotragus moschatus).* von Walther (1964) found that sitatunga cows *(Tragelaphus spekii)* in captivity nursed strange calves either alone or simultaneously with their own.

In the bison *(Bison bison)* cows gave birth when isolated, or in small cow groups (McHugh, 1958), and recognized their own calves. For the first few hours after birth, other cows, even ones with their own calves, would approach and sniff and lick the calf. Calves were occasionally seen suckling the wrong cows. M. Altmann (1961, 1963) found that an aging cow was sometimes left with calves and that unrelated adults as well as the mother would shield a calf from intruders. Domestic cattle seem rather specific to their own calves, as Brownlee (1957) reported that once the calf has been cleaned by the mother it may be difficult to get her to take any other calf.

There is a large volume of work on the specificity of ewes and goats, as the acceptance of the offspring by the mothers is of importance in these largely domesticated animals. Thus L. J. McCann (1956) reported that in mountain sheep *(Ovis canadensis)* ewes usually resort to inaccessible places soon after the lambs are born, and remain close to them for about a week. He and Murie (1944) record that one or more ewes may care for a group of lambs while their mothers are off feeding. Spencer (1943) reported the same behavior in the rocky mountain bighorn sheep *(Ovis canadensis).* Although ewes remaining with groups of lambs have been referred to as "nursemaids," no evidence has been presented that they show any kind of care for such lambs. It would be interesting to know exactly how these groups form. In domestic sheep

(Ovis aries) living in fields, Scott (1945) found that a ewe and lamb tended to be apart from the flock until the lamb was about 10 days old. von Haas (1959) studied a small flock of Barbary sheep *(Ammotragus lervia)* living in an enclosure, and thought that the mothers could not distinguish their own young from those of other females, as very few ewes attempted to ward off a strange lamb, and ewes offered a number of young seemed unable to find their own. It is interesting that he also noted that these sheep are supposed to be solitary under natural conditions.

As in the cervids, the behavior of the preparturient female in going into isolation for the first few days of her offspring's life may enable them both to learn each other's characteristics sufficiently to permit individual recognition on their return to the flock. There is plenty of evidence that ewes, and also nanny goats, recognize their own young. Thus Herscher *et al.* (1963a) noted that both species are consistently violent in rejecting all young other than their own. However, pre-parturient ewes are well known to show interest in lambs, and will persistently try to adopt one (Fraser, 1937), sometimes driving off the lamb's own mother. Such behavior has been seen as early as a fortnight before parturition, and ewes have been seen licking and giving suck to other lambs during the birth of their own lambs.

L.   BEHAVIOR OF FEMALES TOWARD YOUNG: SUMMARY

In the light of this survey it is difficult to make any useful summary of the behavior of females toward young other than their own, both be-cause of the lack of good data on the subject and because very many species are completely unknown in this respect, and also because of the wide range of situations in which the young are reared. Reverting to the classification of the latter introduced at the beginning of this section, it is clear, first, that in situation 1 females range from being completely nonspecific, as in some species of Chiroptera, to being highly specific, as in many ungulates. Second, in situation 2, females of at least some species of several orders (e.g., primates, carnivores, rodents) will attempt to care for the young of other females. There are some minor orders whose "maternal" behavior is little known in this or any other respect.

In many species the mother bears and rears her young in isolation from all female conspecifics, but it cannot be assumed that such animals would not show care for the young given the opportunity.

Not surprisingly, the care shown by females is generally very similar to that shown by the mothers, although some of the behavior shown by her, such as suckling, may be absent. The cases where particular females appear to act as nursemaids to groups of young are of especial interest.

The effects of hormones on the behavior of females in these respects are not well known. Lehrman (1961) and Richards (1967) have reviewed this field. Experiential and other effects are discussed below (Section VII).

IV. SPECIFICITY OF OFFSPRING

Mothers of many and perhaps nearly all species learn the individual characteristics of their offspring, and can differentiate between them and those of other mothers. Eliciting the proper sequence of behavior in the mother undoubtedly depends to a greater or lesser extent on the proper response being given to her by the young. This is obvious in the matter of suckling, but may also be true of other patterns shown by mothers toward their young. The question of how far young recognize their mothers and will or will not respond normally to mothers other than their own is thus an important one. In most cases practically nothing is known about the age at which young can begin to discriminate their mothers from other conspecifics.

It is implicit in the records of females tolerating the suckling attempts of young other than their own (Section III) that young were attempting to suckle animals other than their own mothers. Attempts to suckle from other animals does not necessarily mean, of course, that they cannot distinguish their mothers from other animals, and while young may never attempt to suckle from another animal when their mother is present they may do so in her absence. Further, where several litters are reared in a single den or nest and all suckled by several females, the young are not necessarily incapable of distinguishing between the females, or of distinguishing these from all others.

In cases where the male shares the nest or den with the female it would be interesting to know more about the behavior of the young to each of the parents, and to what extent and in what ways it differs.

The literature at present contributes very little on these questions, and further investigation is needed of the many species in which the young are partly cared for by the male or other females besides the mother. It is particularly urgent in those species in which the control of "maternal behavior" is being investigated using cross-fostering techniques, and in which the full cooperation of the young is assumed.

A. CHIROPTERA

The information in the section on the behavior of females to young (III, C) indicates that the young are nonspecific and will attempt to suckle any female. Gates (1941) for *Nycticeius humeralis* noted that young may even go to a nonlactating female, and Davis *et al.* (1968) found that *Eptesicus fuscus* young tried to grab any adult that came to them. An interesting observation was made by Racey (1969) of a baby *Pipistrellus pipistrellus* which was found still attached to its dead mother in a caged group. Attempts to foster it on a lactating female from another cage (and originally from another colony), failed, because it crawled straight off her, but it immediately took the nipple when put on a female from the same cage and colony as its mother.

B. PRIMATES

This is the group in which perhaps the most interest in young is shown by other animals, and the young of many species accept care from other group members (see Section III, D). However, they recognize their own mothers, perhaps almost from the time they first begin to leave her. Jay (1963) observed that 2-week-old infant langurs *(Presbytis entellus)* readily distinghished their mothers from 15–20 feet, and actively oriented to them. Booth (1962) said that *Cercopithecus* spp. distinguish their own mothers after some weeks, but gave no details. There is some evidence regarding the infant's recognition of their mothers from their response to attempts by animals other than their mothers to hold them. Thus Kaufmann (1966) and Rowell, Hinde, and Spencer-Booth (1964) for free-ranging and captive rhesus monkeys, respectively, noted that infants would refuse to cling to group companions which clasped them. Sugiyama (1965b) recorded that even in the first few days of life infant langurs would resist being taken from their mother by other females, and would squeal and cling to her, having to be removed forcibly, although the mothers were usually quite permissive; both he and Jay (1963) found that infant langurs would try to move away from the females holding them. Kummer (1968) observed that hamadryas baboon infants may try to escape from their male kidnappers.

However, in cases where the infant's mother has been removed, at least in rhesus monkeys, the infant may cling to a foster mother. Spencer-Booth and Hinde (1967) removed the mothers of four 30- to 32-week old rhesus infants, for 6 days, and three of these infants clung to female group companions during the separation period, and for a much higher proportion of the time watched than before separation. Whereas these animals were never observed to take the nipple, E. W. Hansen (1966) fostered a 37- and a 78-day-old infant on females and both infants took the nipple.

Human babies do not at first distinguish between caretakers, but come to distinguish individuals later. Ambrose (1961, 1963) studied the smiling response of babies. The response to the experimenter's face increased until about 15 weeks of age and was followed by a rather sudden drop in responsiveness. This was thought to be due to the discrimination of the mother's face from those of other people beginning at this time, and possibly to the intervention of fear responses to strangers. The development of attachment to the mother in the Ganda people has been described by Ainsworth (1963), and Schaffer (1963) has also discussed the formation of attachments by babies. Ainsworth describes a case of apparently successful multiple mothering in which the child nevertheless distinguished and preferred its true mother in times of stress.

C. Rodents

Section III, E gives references to communal rearing of litters and acceptance of foster young by mothers which implies that the young are nonspecific, e.g., Goodpaster and Hoffmeister (1954) for *Peromyscus nuttalli,* King (1955) for *Cynomys ludovicianus.* Denenberg *et al.* (1962) cross-fostered rat pups with varying success depending on the conditions of the experiment, and Beniest-Noirot (1962) and Rowell (1960) did the same with mice and hamsters, respectively (see Section VII).

Among the hystricomorphs Roth-Kolar (1957) notes that *Dasyprocta aguti* in captivity could recognize their own mothers within 3 days, having approached all conspecifics without hesitation until then, and King (1956) found that under seminatural conditions young guinea pigs *(Cavia porcellus)* followed other members of the colony, but to a lesser extent than they did their own mothers. Kunkel and Kunkel (1964) found that in large captive groups a female might be followed by a number of young of widely different ages at the same time.

Despite these observations, which seem to suggest that the altricial young of this group generally accept any mother who is willing to accept them, it remains possible that they nevertheless distinguish between their own and other mothers. Rowell (1960) in a careful study of hamsters found that once a pup's eyes had opened its reactions to an unfamiliar female were different from those to their own mother. It might be instructive to look more closely at the pup's responses in cross-fostering experiments; the blame for nonsurvival of foster pups should perhaps not be laid entirely at the mother's door. Frank (1952) noted that young of *Microtus arvalis* in captivity resist being retrieved by animals other than their own mothers.

D. Cetacea

Whereas young seem normally to suckle only from their own mother (see Section III, G), an orphan *Tursiops truncatus* put into a tank containing a captive group immediately tried to suckle two pregnant females (McBride and Kritzler, 1951).

E. Carnivora

1. *Pinnipedia*

As shown above, the young tend to be rejected by females other than their own mothers. Bartholomew (1952) noted that young *Mirounga angustrirostris* attempted to nurse any adults, and Fogden (1968) found that pups would suckle, or solicit suckling from, several females within a few minutes of each other. He and E. A. Smith (1968) also found that gray seal *(Halichoerus grypus)* pups would suckle from more than one female. Laws (1956) thought that young *M. leonina* did not recognize their own mothers, as a hungry pup on its own would make for the nearest seal. It is not clear whether they would make for their mother if put into a situation where they had a choice.

2. *Fissipedia*

In species where litters are reared communally, young appear not to discriminate between females (e.g., *Lycaon pictus*). In such cases one should beware of inferring that the young cannot distinguish their own mothers, or do not prefer her to others if given a choice.

A little experimental work has been done with domestic cats and dogs, and both species apparently can be fostered (cats: Ewer 1961, Schneirla *et al.,* 1963; dogs: Menzel and Menzel 1953). Schneirla *et al.* (1963) found that kittens can distinguish their own mothers shortly after birth, and take a little time to accept a foster mother. They studied the effects of isolating kittens in brooders for different periods at different ages on their response on return to their mothers. It was found that although suckling appeared in all groups, it did not appear immediately in any individuals, taking up to 3 hours in kittens isolated from birth to Day 7, 20 hours in the Day 6–23 group, and 15 hours in the Day 18–33 group. Not only do kittens recognize their mothers, but they may be specific to a particular teat, or pair of teats, although Schneirla *et al.* (1963) and Schneirla and Rosenblatt (1961) noted that some kittens did not acquire a nipple specificity. Rosenblatt and Schneirla (1962) found the beginnings of nipple preference in some kittens by the first day. Ewer (1959, 1961) found that once a nipple preference had been established a kitten's response became restricted to one teat, which it might fight for if it found it occupied. A kitten put with another mother and litter would reject all teats, including the anatomically "correct" one.

Domestic dogs apparently do not usually exhibit nipple preferences (Fuller and Dubuis, 1962) although Rheingold (1963) recorded a nipple preference in one litter out of five observed.

F. PROBOSCIDEA

An African elephant calf was observed to suckle from a female with a calf of her own despite the presence of its own mother (Douglas-Hamilton, 1969).

G. UNGULATES

1. Perissodactyla

There is very little information on this group, but it appears that foals will attempt to suckle mares other than their own mothers [see Section III, K; Wackernagel (1965) for zebras, and Tyler (1969) for ponies] and are thus to some extent nonspecific to their mothers, although they recognize them.

2. Artiodactyla

While the behavior of the mothers is usually to reject violently all young other than their own, frequently the young approach other mothers (see Section III, K). Thus Herscher et al. (1963a) and Collias (1956) reported that hungry goat kids will attempt to nurse any dam. Katz (1949) studied a small herd of Barbary sheep *(Ammotragus lervia)* and found that one lamb followed two ewes alternately.

McHugh (1958) observed that buffalo *(Bison bison)* calves seem to recognize their own mothers and moved directly to them. Hafez and Schein (1962) note that in domestic cattle, once a strong pair-bond relationship has been formed, the calf is very distressed by separation from the mother. Estes (1963) observed in wildebeest that the calves cannot initially distinguish their own mothers, but can single them out after about a day. Cuneo (1965) noted that in captive klipspringer antelope *(Oreotragus oreotragus)* after about 3 months the young would follow any of the females in the group.

Hafez and Scott (1962) quoted Cowlishaw as saying that with twin lambs in domestic sheep, each has its "own" teat, although singletons suckle from both nipples.

In *Cervus elephas* Burckhardt (1958) observed that young seemed to recognize their own mothers and did not attempt to suckle strange ones, unlike young chamois *(Rupicapra rupicapra)*. M. Altmann (1963) discussed the development of the following response of calves of elk *(Cervus canadensis)* and moose *(Alces alces)*. The former may attempt to suckle strange cows; it is not clear whether moose calves would have done so if the mothers had not driven other animals away. Harper *et al.* (1967) described Roosevelt elk *(C. c. roosevelti)* calves as leaving the calf group when their mother called them (implying a response specific to their own mother) and that when the mother rejected them at the end of the nursing period they returned to the calf-group with the baby sitter. Pruitt (1960) describes how caribou *(Rangifer arcticus)* calves followed any cows that bobbed their heads at them, for an unknown period after birth. Lent (1966) observed that caribou calves showed little discrimination for the first day, and even at a later stage would attempt to associate with cows they recognized as not being their own mothers.

In the domestic pig the young apparently become specific to a particular nipple, and Hafez *et al.* (1962) record how this is gradually established by the litter. However, it seems that they are not specific to their own mothers, as these authors note that foster young will attempt to take up the nipple *position* that they had on their own mothers, but experiments giving choice of mothers do not seem to have been done. G. McBride (1963) has also discussed the establishment of nipple preferences and observes that a piglet which has been driven off its "own" nipple may refuse to accept another one and eventually starve. This seems to represent a high degree of specificity in the young.

Koford (1957) found that although vicuna *(Vicugna vicugna)* young would attempt to nose at adults other than their mothers, they seldom did so once they were several weeks old. Pilters (1954) found that zoo-living young were generally looked after in isolation from the rest of the group for the first week of life, but that if a second mare was present the young would go to her as often as to their own mother. At the first meeting with the herd llama foals were recorded as following other conspecifics of both sexes, and camel foals as running from one mare to another. Gauthier-Pilters (1959) observed that young dromedaries *(Camelus dromedarius)* recognized their own mothers.

H. SPECIFICITY OF OFFSPRING: SUMMARY

The information is too meager for very much to be said in summary of this section. Clearly the young of several orders will actively attempt to suckle females other than the mother (e.g., Chiroptera, pinnipede Carnivora, Ungulates). Those of other orders will suckle from females other than the mother that happen to be present in the nest or den (e.g., some rodents, some fissipede Carnivora). However, the young of some orders clearly discriminate their mothers from other conspecifics (e.g., primates) since they may avoid the attentions of such animals. Furthermore, there is clear evidence from some of the species in which the young will suckle females other than their mothers that they nevertheless recognize their own mothers (e.g., golden hamsters, some ungulates). More extensive studies might well reveal that individual recognition of mothers is much more widespread than it appears at present.

V. THE FORMATION OF THE MOTHER-INFANT RELATIONSHIP

Undoubtedly the behavior of both mother and young are important in forming and maintaining this relationship. It is not proposed to discuss here the nature of the attachment process, which has often been referred to as "imprinting" or discussed in that context (e.g., Klopfer *et al.* 1964, for goats; Espmark, 1964b, for reindeer; Darling, 1938, for red deer; Hess, 1959, for guinea pigs; Rosenblatt *et al.,* 1961, for cats). The use of this term has been discussed by Bateson (1966, 1969) and Hinde (1966). The use of the term by Herscher *et al.* (1958) and Scott (1960) in relation to the attachment process has been discussed by Denenberg (1962).

It is important, however, to be clear about the precise conditions under which the attachment is formed and maintained. Cairns (1966) has stressed the importance of the isolation of mother and young, and noted that in animals bearing precocial young such as ungulates this isolation is achieved by the behavior of the mother in removing herself from the herd at parturition. However, while in some species (e.g., moose) the young are born in isolation and kept away from the herd for some time afterward (e.g., 7 days, mountain sheep; 1–2 weeks, elk) in other cases the herd may be rejoined as soon as the young can walk (e.g., ponies) and in yet others they may be born in a maternal group (e.g., wildebeest). This situation is also true of some Pinnipedia and some Chiroptera, and although the young are rather helpless at first and the mother may remain close to her offspring as in Pinnipedia, or may carry it, as in Chiroptera, for a while after parturition, the relationship a little later on has to be maintained in the face of tremendous opportunity for interaction with other animals, both for the mother and for the young. This is also true of the higher primates in which the young are relatively precocious and are also surrounded by group companions from birth. The situation here is even more complex as nonbreeding females and males are not usually excluded from the group. Cairns stressed that a further factor in the maintenance of the mother-infant relationship in ungulates is the rejection of alien young by females. However in approaching alien females, the young are showing that their attachment to their mothers allows them to initiate interactions with others, even to suckling from them if they are permitted to do so. Furthermore females are maintaining attachments to their own young in the face of these approaches, or at least failing to form attachments to alien young. In the case of primate females these comments apply in reverse as these usually initiate the interactions with young other than their own and there may be rejection by the young.

Cairns (1966) thought that in species bearing relatively helpless young the formation of attachments to animals other than the mother was prevented by the incompetence of the young. Perhaps a somewhat different viewpoint should be put on this. In many cases, such young are exposed while still helpless to interactions with adults (e.g., African hunting dogs). In such cases, although it may not be necessary for them to form an attachment to the mother alone, neither is it known to what extent they discriminate her from the others. The same is true of species in which the male helps the female to rear the litter (e.g., coyotes) or is at least present in the nest (e.g., *Peromyscus maniculatus*).

VI. MECHANISMS OF MOTHER-INFANT RECOGNITION

With the exception of ungulates, in which it may be of practical importance, the factors that enable mothers to distinguish their own offspring from those of other mothers and vice versa have not been very carefully investigated. In any case they are probably complex.

Possible factors have been suggested in some cases from observations of the actual behavior which preceded acceptance or rejection of mother or young, such as the mother sniffing the young first. Such observations can do no more than act as guides for more precise work. Furthermore, the response of a female (known to be able to distinguish her own young) to a signal in a particular modality from any young of the species, does not necessarily mean that she cannot distinguish such a signal by her own young from that by other young. Beach and Jaynes (1956c) studied the sensory cues involved in a female rat's response to her own young, and concluded that they were probably multisensory.

The modalities which are most often invoked as providing the mechanism for individual recognition are smell and hearing. The ability of some mammals, such as dogs, to distinguish smells of different people is well known; and Thorpe (1968) has discussed the parameters available for recognition in a complex sound, and stressed the suitability of this modality for individual recognition between conspecifics.

A. CHIROPTERA

In this group it is thought that recognition may be by smell (e.g., Pearson *et al.*, 1952), or vocalizations, particularly ultrasounds (Brosset, 1966).

Kleiman (1969b) reached the former conclusion after studying the noctule *(Nyctalus noctula)*, pipistrelle *(Pipistrellus pipistrellus)* and serotine *(Eptesicus serotinus)* although she thought that there might also be individual differences in vocalizations. The muzzle gland secretion of the mother may provide the mother and young with the same individual scent in the serotine (Kleiman, 1969b) and long-eared bats *Plecotus auritus* and *P. austriacus* (Stebbings, 1966). It may be that colony or group odor, as opposed to individual odor, can be distinguished by the young (see Racey, 1969, Section IV, A for pipistrelles) although presumably mixing of group members does not normally occur. Davis *et al.* (1968) think that squeaking of the young of the brown bat *(Eptesicus fuscus)* helps the mother to locate them.

B. PRIMATES

Individual recognition has not been investigated in any detail in this order, in which it is very highly developed. Carpenter (1934) thought that in howler monkeys *(Alouatta palliata)* mother-infant recognition may be by means of the vocalizations and probably other cues such as smell. Moynihan (1964) stresses the importance of vocal communication in the nocturnal *Aotus trivirgatus*, but has no actual evidence of individual recognition by this means.

C. RODENTIA

Rats were thought to recognize their own young by smell by Beach and Jaynes (1956a) but Beach and Jaynes (1956b) found that olfaction was not essential for retrieving to occur, and thought that multisensory cues are normally involved in promoting retrieving.

D. CARNIVORA

J. E. Hamilton (1934) thought that the southern sea lion *(Otaria byronica)* mother recognized her own pup by voice. Eibl-Eibesfeldt (1955) noted that Galapagos sea lion *(Zalophus wollebacki)* mothers responded only to calls of their own young, but apparently also recognized them by sniffing at them. Bonnot (1928) saw lactating females approach a distressed and calling pup and leave it after sniffing it. He thought the pup recognized the individual call of its mother, but the evidence is not very conclusive. He also thought calls were important for individual recognition between Steller sea lion cows and calves *(Eumetopias stelleri)*.

Laws (1956) thought that female elephant seals *(Mirounga leonina)* recognized their own young by smell. Klopfer and Gilbert (1966) disagree with Carrick *et al.* (1962), who thought that a pup would be rejected by a female other than its mother if it vocalized. They found that all females tended to respond to the bawling of an isolated pup, but that they apparently distinguished their own by some other factor.

E. UNGULATES

Domestic piglets are known to recognize their "own" nipples (see Section IV, G), but the means by which they do so is not clear. Hafez *et al.* (1962) thought that recognition is facilitated visually by perception of the mother's body and udder conformation, that tactile cues are of secondary importance, and that auditory and olfactory ones are not important. Donald (1937) found that more errors in locating the correct nipple occurred at the center of the udder than either end, which supports the suggestion of the importance of spatial orientation, and he also found that the suckling order did not change when the udder was coated with mud or a scented substance. G. McBride (1963) thought that sight, smell, and recognition of neighbors were important factors.

Hafez *et al.* (1962) say that a sow recognizes her own litter by olfactory cues, and that smearing a sow's litter with her excreta may help to foster them on her, as may confining them with her own piglets for 2–3 hours.

Licking the newborn does not always seem to be important, as vicuna females *(Vicugna vicugna)* do not groom their offspring (Koford, 1957) although smell may be important in recognition by both sides. Hodge (1946) describes the fostering of vicuna young on alpaca mothers *(Lama pacos)* by draping the vicuna in the skin of the alpaca's newborn young.

Gauthier-Pilters (1959) noted that dromedaries recognize their mothers mainly by their voices. Pilters (1954) thought that camel foals recognize their mothers by smell very early, but hardly at all by sight, whereas he thought that llama young recognized their mothers by sight.

The Cervidae are another group in which smell seems to be important. Thus Altmann (1952) observed that in elk *(Cervus canadensis)* several cows would respond to a bleating calf, and that individual recognition seemed to be by smell. However, Lent (1966) thought that caribou *(Rangifer arcticus)* cows and calves recognized one another individually by their bleating. Espmark (1968) is quoted by Thorpe (1968) as thinking that young reindeer recognize their mothers even if they are not visible, presumably by sound.

In the black-tailed deer *(Odocoileus hemionus)* recognition seems to be by scent as young who approached females would be smelt or licked before being rejected or accepted (Golley, 1957).

McHugh (1958) said that buffalo *(Bison bison)* seemed to identify their mothers largely by sight, and perhaps sometimes by scent and sound. The cows apparently recognized their calves by sniffing them, and also sometimes by sight or sound. In domestic cattle Brownlee (1957) stressed the importance of licking the newborn by the mother for establishing the relationship between them. Murie (1944) noted that for Dall sheep *(Ovis dalli)* the lambs may be able to recognize the bleating of their own ewes.

Most of the work has concentrated on the recognition of domestic goats and sheep by their mothers. Collias (1956) noted that a mother goat or sheep may call in response to the bleating of any young of the species (or of the other of these two species), although she may later re-buff its approach. Hafez and Scott (1962) said that the development of the relationship between mother and young in goats and sheep depends on constant contact. The period immediately after birth is clearly of great importance as Collias (1956) found that if the lamb or kid was re-moved from the mother at or just after birth and kept away for 2 hours or more, the mother was likely to reject it on return. Herscher *et al.* have also found that the first few minutes (1958) after birth are im-portant. Kids were removed from their mothers for ½–1 hour at 5–10 minutes postpartum. The mothers were tested in isolation from the flock with their own and strange kids 2–3 months later. Half of them butted alien kids, not their own, but nursed indiscriminately, and half refused to nurse at all. Control animals which had no separation experience did not nurse indiscriminately. The authors suggested that the 5–10 minutes is sufficient to establish the identity of her own kid to the mother, but that a longer period is necessary for the development of an individual specific nursing bond. Licking the young and the birth membranes may facilitate, but are not essential to, the formation of the bond between mother and young. Collias (1956) found that if a newborn lamb or kid was washed with detergent it would still be accepted by its mother. They also successfully fostered a newborn kid on a mother whose own young had been removed after she had licked it.

Herscher *et al.* (1963) thought that in sheep and goats sniffing was the main mechanism for recognition of young although vision may play a role. However Klopfer *et al.* (1964) concluded that vision was not important in goats. These authors thought that the immediate response was to a highly labile factor in the birth fluids, and that the effluvia from the gut of the newborn might have some specific factors in common with the fluids. They found that newborn kids which had not been licked or nuzzled were subsequently rejected after a 1–3 hour separation, whereas those which had been with their mothers for 5 minutes from the first licking or nuzzling were accepted. However, they found that although licking of a foster kid may be stimulated by rubbing it with a female's birth membranes, this does not lead to nursing. Klopfer and Gamble (1966) found that olfactory impairment during parturition did not lead to rejection behavior, but abolished alien-own young discrimination. However, olfactory impairment at the time of testing the response to young after separation could produce rejection of own young. Fraser (1937) noted that once a ewe has licked her lamb she is most unlikely to leave it, and draping a foster lamb in the skin of a mother's dead lamb is a common but apparently not very successful technique for promoting adoption (Fraser, 1937; Lamond, 1949). The latter recommends spraying with the mother's milk. Moore and Moore (1960) suggested that in goats the environment of the mother while separated from her kid is important in determining her behavior on its return. Mothers which had mingled with the flock rejected young after short periods of separation, but accepted them if they had been isolated during such periods.

F.   Mechanisms of Mother-Infant Recognition: Summary

Many of the views as to the modality by which the mothers and young recognize one another are based on rather incidental observations of the behavior leading to their approach to one another, or of their behavior when they contact one another. With the exception of domestic ungulates such as sheep and goats, and the laboratory rat, little systematic experimental work has been done.

VII.   Factors Affecting Whether or Not Care Will Be Shown

The mechanisms involved in recognition between mother and offspring have been discussed above, including those which seem to be important in the period immediately after birth. This refers to the establishment of the individual mother-infant bond. A different issue concerns the factor(s) that determine whether a female, when exposed to young, will show "maternal behavior." The behavior of the true mother in this respect has been omitted; for instance, the effect of the number of previous litters on "maternal behavior" is strictly speaking concerned with the effect of a female's experience with her own young on her behavior to her own young, and is not relevant here. The discussion is confined to the experiential and environmental factors which seem to affect the responses of females, other than the mother, to young. The work concerned with the hormonal factors affecting such response has been reviewed elsewhere (Lehrman, 1961; Richards, 1967) and is not included. Of course, environmental and experiential factors may not be affecting the behavior directly, but through the endocrine system.

The most important variables that may affect a female's response to young are the age of the young, whether or not she has borne young and, if nulliparous, whether she has had previous experience of young. Of course, the stimuli received from the young affect her behavior too, and thus not only will their age be important but also the nature of their response to the female. These factors are all concerned with a general response to young of the species, but the presence of a female's own young may affect her response to other young. In this case the female is already responding to her own young, and the question is whether she will respond to any others. Acceptance of young other than her own by a lacating female does not necessarily mean that she cannot distinguish her own.

A. AGE OF YOUNG

Obviously there must be an age for the young of all species at which they no longer elicit maternal care. However, the time course of changes with age of young of the responsiveness to them has been investigated in only a few species.

1. Primates

The interest in infants shown by primate females changes with the age of the infant (see chapters in DeVore, 1965; Hopf, 1967). It has also been noted in baboons that newborn young are not an immediate source of interest (Hall and DeVore, 1965), but DeVore (1963) for baboons and other authors for other species (e.g., Jay, 1962, 1963, for langurs) have described interest during and immediately after the birth. Rowell, Hinde, and Spencer-Booth (1964) found that in captive rhesus monkeys interest in infants by females other than the mother was often evident within a few hours of their birth, and the type of behavior shown by such females changed with the infant's age. Spencer-Booth (1968) found that the amount of approaching and looking at the infant was highest in the first six weeks, and touching and pulling at infants in weeks seven to twelve; cuddling was highest in weeks seven to eighteen, whereas the amount of grooming of infants reached a peak in weeks thirteen to eighteen. The type of response that a female can show to an infant is determined to a large extent by its mother, and the changes with age of infant in the mother-infant relationship (see Hinde and Spencer-Booth, 1967) may account for the time course of the responses outlined above.

However, the cessation of interest by females has been linked in a number of primate species with rather a marked change in the appearance of the infant, such as in coat color. Jay (1962) has remarked that in all Old World monkeys for which information is available the natal coat color is different from that of the adult. Rowell *et al.* (1964) noted that the heads of newborn rhesus have two tufts of hair, separated by a center parting, and that when the natal coat is lost and replaced by a thicker one, this disappears, changing the shape of the head. This process begins at about 12 weeks of age. Jay (1962) said that in langurs the interest of adult females in infants other than their own begins to wane when the brown natal color begins to turn to white. Kummer (1968) emphasizes that in hamadryas baboons it is the younger black infants that attract the attentions of adults rather than the older ones whose coats have turned brown. Hall and DeVore (1965) found that in the chacma baboon the adults considerably reduce their attention to infants in the fourth to sixth months, the age at which their coats are changing from black to brown.

2. Rodentia

Rowell (1960) presented lactating female hamsters with foster pups of various ages and found the highest percentage of acceptance with pups aged 7–10 days. Richards (1966b) did similar experiments using virgin females and found that the most effective age for acceptance was 6–10 days, although even these were attacked initially. In laboratory mice the tendency of naive virgin females to show retrieving, licking, nest-building and lactation position responses decreases with increasing age of the young (Noirot, 1964a). Retrieving and nest-building showed a relatively sudden drop on Day 13, and Noirot has suggested that this could be associated with the cessation of ultrasonic cries from the pups at this age. Thus in these two species there is a marked difference in the age at which pups provide the most effective stimuli for these responses from the females, and Richards (1966b, 1967) has discussed the possibility that this can be accounted for by a recently evolved reduction in the gestation period for hamsters.

Yardeni-Yaron (1952) noted that in captive Levante voles *(Microtus guentheri)* males and females with young would most readily accept strange young which approached their own in age, though surprisingly age of young apparently was not a factor affecting acceptance of young by voles which did not have young of their own.

Working with laboratory rats Denenberg *et al.* (1962) found that young fostered immediately after birth were less likely to survive than young fostered 12 hours after birth, and suggested that the natural mother's behavior during the first hours of life are critical for survival of the young. However, the reasons for this effect are not at all clear as the behavior of the animals was not observed.

Seward and Seward (1940), using a hurdle box to test female guinea pigs, thought that the age of the young was not important in affecting the responsiveness.

3. Ungulates

Estes (1963) found that whereas wildebeeste cows were intolerant of new calves, they were very tolerant of those a few days older, although probably rarely to the extent of letting them suckle. He noted that the hostile response corresponded closely to the stage during which the calf is unable to distinguish its own mother.

B. PRESENCE OR NOT OF "OWN YOUNG"

1. Primates

Kaufmann (1966) found that in rhesus monkeys mature females holding an infant of their own would sometimes hold another (motherless) infant simultaneously. Spencer-Booth (1968) found that among the females which had born live young, those of them which had similar-aged young of their own interacted with other infants least, although they were more likely to hit infants than other categories of females. Sugiyama (1967) noted that in langurs *(Presbytis entellus)* it was only the nonlactating females which held young, but Jay (1962) saw both lactating and nonlactating females do so.

2. Rodentia

Rosenblatt (1965) described the effect in rats of the presence or not of the female's own litter on her responsiveness to test pups. When mothers were rearing their own litters they retrieved and nursed test pups and did not attack them. When mothers which had had their own litters removed at parturition (but after having licked them) were tested with pups they were clearly less "maternal" than those with their own litters. Thus at the first testing on the sixth day and at three subsequent weekly tests thereafter a much lower percentage of the experimental group than of the control group retrieved and nursed the pups. Nest-building also declined markedly when the pups were removed at parturition.

Presence of young may also have decremental effects. Noirot (1964c) compared female laboratory mice living continuously with their own litter with virgin females which were presented with a test litter for twenty daily 5-minute periods. Licking and lactation position were less frequent in the former group than the latter, although there was no difference for retrieving and nest-building. Thus some kind of habituation to the young may have resulted from continuous living with them. [Note that Beniest-Noirot (1958) found no difference between virgin and newly parturient females in their responsiveness to young pups.]

3. Ungulates

Lent (1966) observed that in caribou intolerance of strange calves did not seem to depend on whether or not the cows had their own calves beside them. However, Pruitt (1960) found that cows which had lost their young would attempt to kidnap the young belonging to other cows. Estes (1963) found that wildebeeste cows were intolerant of new calves whether or not they had calves of their own, but that it was the mothers with their newborn calves that were the most intolerant. In domestic sheep the attempted kidnapping of lambs ceases immediately a ewe's own lamb has been born (Fraser, 1937); since the behavior may be shown by a ewe in successive breeding seasons, this change seems to be an effect of presence of own young, rather than a change in responsiveness with parity (although note that if her lamb is removed the ewe will still refuse foster lambs).

D. The Effect of the "Correctness" of the Young's Response

In a number of cases cross-species fostering has proved perfectly possible (e.g., Kahmann and von Frisch, 1952, for several species of rodents; Cox, 1965, for note of *Mus musculus* rearing *Apodemus sylvaticus;* Ackert, 1910, for *Peromyscus maniculatus* rearing *Mus musculus;* Herter and Herter, 1955, for polecat adopting domestic kittens; Loisel, 1906, for domestic dog accepting rabbits). Thus in some cases acceptance by a female does not necessarily depend on her receiving species specific responses from the young.

However cross-fostering is not always completely satisfactory, or even possible. Denenberg *et al.* (1966) fostered mouse pups on rats and found that they were accepted, although there was a higher mortality than when they were reared by their own mothers. They interpreted this result as showing that the retrieving response in the rat is a reaction to a "young" stimulus, rather than to a stimulus having specific properties such as a certain size or shape. Beach and Jaynes (1956c) found that rats would not retrieve young guinea pigs. Ressler (1962) cross-fostered mice of two strains, and concluded that the behavior of the young, which differed with the strain, was an environmental factor affecting parental behavior.

Two other cases where the response of the female seems to be affected by the response of the young are worth mentioning. Berlioz (1933) noted that buffalo cows *(Bos bison)* bred to domestic bulls *(B. taurus)* often failed to suckle the hybrid calves, and Hafez and Schein (1962) thought that this might "reflect impaired recognition of the calves." Harlow *et al.* (1963) found that a rhesus monkey female would not adopt an infant which would not cling, and tended to avoid it. However, a rhesus monkey will sometimes carry a dead infant of her own (Spencer-Booth, personal observation) so the effect of an abnormality in a true offspring may be a different issue from the effect in a foster one.

Finally, the behavior of the human child has been shown to affect the behavior shown toward it by adult caretakers. The literature has been critically reviewed by Bell (1964a,b), who emphasizes the importance of taking the behavior of both partners into account when assessing mother-infant relationships.

D. The Effect of Previous Experience with Young

The experience of having borne young of her own may affect the behavior which a female shows to the young of other mothers, and there may also be a change in responsiveness of nulliparous females as a result of experience with the young of other mothers. Furthermore there are two kinds of effects. First, the response to the stimulus shown may be qualitatively quite different as a result of experience of the same stimulus situation, of part of it or of a quite different stimulus. Second, there may be a quantitative change in sensitivity, which may be shown in several ways. The form of the change in any particular case may depend simply on the way the effect is measured, or there may be a real difference between cases in the nature of the change. Thus the experience may change the time spent responding or the latency to respond, or there may be a change in the proportion of the animals tested which respond (i.e., while some were above/below the response threshold before the experience, others crossed the threshold as a result of the experience). Another effect on the animal may be that it requires either a less or more complete stimulus situation for the response to be shown as a result of the experience.

The results for the few species which have been studied are arranged according to order, and whether the effect is of experience with "own young" or "other young." The possible nature of the effects is discussed thereafter.

1. Experience with "Own Young"

a. Primates. There is little information for primates. Jay (1962) found that in langurs *(Presbytis entellus),* although both subadult and adult females showed intense interest in babies, the former seldom held them. However, this may not be a result of bearing young and may not even reflect a difference in interest or the degree of "maternal-ness" so much as a difference in opportunity. Female howler monkeys without babies were especially interested in infants (Carpenter, 1965), but it was not stated, and probably was not known, whether these females had ever had babies.

DeVore (1963) found that in baboons the interest was greatest among the juvenile and subadult females. Kaufmann (1966) found that in free-ranging rhesus monkeys immature females as well as adults, with or without babies of their own, were all interested in infants, but especially the newly mature females, which presumably had not had babies of their own. Spencer-Booth (1968) found that in small captive groups of rhesus monkeys females which had never born live young showed more touching, cuddling, and grooming of infants than those that had done so. Among those which had not born live young, those about two years old when the baby was born showed most of these kinds of interest in it.

Cross and Harlow (1963) tested rhesus females in Butler boxes and found that multiparous nonlactating females preferred viewing an infant monkey to a juvenile one, and those whose babies had just been removed at a few days old also preferred viewing babies. Nulliparous females showed no preference for either. However this experiment is not very informative as there can be no estimate of what sort of behavior the females would have shown to either young or juveniles had they had access to them.

b. Rodentia. Brown (1953) investigated the behavior of mice *(Mus musculus)* in relatively large enclosures and found mild interest shown by nulliparous females, compared with the much more intense interest shown by lactating females. However, as in the case of the results of Richards (1966a) in which lactating female hamsters accepted test pups and virgin females killed and ate them, the effects of experience and hormonal state are confounded. This difficulty was overcome in experiments by Noirot (1964a), who found that although naive females of laboratory mice did show retrieving, licking, nest-building and lactation position to pups (see also Beniest-Noirot, 1958), previous experience of test pups increased responsiveness to them. There is also evidence that experience with young exerts negative effects (see Section VII, B; Noirot, 1964c).

Cosnier (1963) found that in rats multiparous nonlactating females had a consistently higher score for maternal behavior than nulliparous females. Wiesner and Sheard (1933) found that a higher percentage of rats would retrieve test young after they had had young of their own than before. Moltz and Wiener (1966) found that when females had their own pups delivered by cesarean, fewer primiparous than multiparous females would accept newborn foster pups.

Richards (1967) has reviewed the literature on rodents.

c. Ungulates. In the ungulates there is no real information on this factor, as nonpregnant yearling females are normally excluded from the parturient group. Lent (1966) described the harassment by yearling female caribou of the first cow of the season to calve, but it is not clear that they were interested in the calf: possibly they were attempting to suckle, which occasionally occurs in yearling ungulates.

2. Experience with "Other Young"

a. Primates. In rhesus monkeys the effect of exposure to infants remains to be investigated, although some effects of having borne young and of the presence of young are known (Spencer-Booth, 1968).

b. Rodentia. The effect of experience of young on the behavior of females toward them has been extensively studied in laboratory mice. Noirot (1964a) exposed virgin females to a series of test pups growing steadily older, and assessed them for retrieving, licking, nest-building, and lactation position. Compared with totally naive females tested with each age of pup, they responded more to all except very young pups. Thus exposure to very effective stimuli (i.e., young pups) increased the tendency to respond to less effective ones (i.e., older pups). However there can also be decremental effects. Noirot (1965) measured the frequency of responses in naive female mice with closely successive presentations of pups, and found that while retrieving remained at a high level, licking and nest-building fell off (but had recovered 24 hours later), and lactation position increased during the early presentations and then remained at a high level. In further experiments (Noirot, 1964b) previous exposure to a very young pup was found to induce these responses by naive females to a drowned pup. Furthermore this effect, at least on retrieving, did not depend on the actual performance of the response; exposure to the stimulus was sufficient, Noirot (1968) found that virgin females which had been in a colony where they could hear the ultrasounds produced by pups, were more responsive to 1-day-old live pups than to those reared in isolation from pups. Indeed, even exposure to the ultrasounds made by a pup enclosed in a tin box was sufficient to increase the percentage of naive females which retrieved a dead pup (Noirot, 1969), although it is possible that the females could also smell the pup in the box and that this was affecting their response.

Golden hamsters seem to show some of the same kinds of effects: Richards (1966a) found that pregnant females first attacked test pups, but later showed maternal responses to them. Noirot and Richards (1966) presented naive females once with either a "strong" stimulus (9-day-old pup) or a "weak" one (1-day-old pup) and found that they subsequently showed more "maternal" behavior to 5-day-old pups than did totally naive females. However, the latency to attack was decreased after initial contact with 1-day-old pups and increased after contact with 9-day-old pups. Females presented twice with 5-day-old pups showed a less marked increase in "maternal" responses, and no change in latency to attack.

Lott and Fuchs (1962) tested the effect of living with pups for 5 days on the response of virgin female rats and found that retrieving was not elicited by this sensitization period. Rosenblatt (1967), however, found that if continuous exposure to 5- to 10-day-old pups was prolonged beyond 5 days, retrieving, licking, crouching, and nest-building were induced, and by 10–15 days almost all animals responded thus.

3. Experience with Young: Discussion

While nothing is known of the physiological mechanisms mediating the effects of experience described above, it may be helpful to consider the results in terms of the possible kinds of effects (in a descriptive sense) that they are producing. In hamsters the effect of the experience with young on the females is to change their response (a) to a constant stimulus in the case of the pregnant females which first attacked and then were "maternal" to test pups (Richards, 1966a), or (b) to a similar stimulus, namely a pup of a slightly different age (Noirot and Richards, 1966).

A somewhat different situation is that of mice in which exposure to one stimulus, the ultrasounds of a pup, increased the number of females responding to another stimulus, a drowned pup (Noirot, 1969). Since there were probably no stimulus characters in common between the two situations, the effect cannot be ascribed to conditioning: Noirot describes it as "priming."

In mice and hamsters that have not had young, exposure increases responsiveness to pups quantitatively, and in mice also increases the response to a dead pup, which is a less complete stimulus than a live one (Noirot, 1964b). In the case of rhesus monkeys bearing live young it may be that "other young" now represent a less complete stimulus than "own young," and so are responded to less (Spencer-Booth, 1968). The behavior of ewes in trying to kidnap lambs until they have borne their own (see Section VII, B) may be a similar phenomenon although here the effect is a short-term one, as they may repeat the behavior in successive breeding seasons.

E. Factors Affecting Whether or Not Care Will be Shown: Summary

Clearly there are many factors besides the female's hormonal state which affect the nature of her response to young. The effects of the age of young and the "correctness" of the young's response to the female are effects of the properties of the stimulus situation and are important in primates and rodents and probably in other orders as well. The presence of a female's own young is also important, and the effects may be incremental (e.g., rats) or decremental (e.g., rhesus monkeys, ungulates) on the female's acceptance of strange young. Previous experience with young is also important and may affect the female's responsiveness in various ways. The effect may depend on whether the experience was with her own young or with young of other females; and again such experience may make them more accepting of strange young (e.g., rats, Mus musculus) or less so (e.g., some primates).

VIII. Discussion

This assembly of observations on the behavior in mammals of animals other than the true mother toward conspecific young cannot shed light

on the control of the behavior which the true mother shows. Nevertheless, four main points stand out clearly. First, the gaps in our knowledge
of the normal behavior of mammals toward young are enormous:
whole orders are virtually unknown in this respect. Second, it is clear
that the term "maternal behavior" is rather useless unless its limitations
in any particular case are clearly defined: for all but very broad defini-
tions careful investigation of the behavior of animals other than the
mother must have been undertaken. Third, it must not be forgotten that
the mother-young relationship is a two-sided one in which each partner
affects the other. Finally, theories of the development of this relationship must either remain specific to particular rearing conditions, or if
more generalized take adequate account of their variety.

Acknowledgments

I am grateful to the Medical Research Council for their support while this work was
written, and to Professor Hinde for his advice and encouragement.

References

Ackert, J. E. 1910. Some observations upon a family of white-footed mice. *Nature-Study
Rev.* **6**, 137–140.

Ainsworth, M. D. 1963. The development of infant-mother interaction among the Ganda.
In "Determinants of Infant Behavior" (B. M. Foss, ed.), Vol. 2, pp. 67–104. Methuen,
London.

Alcorn, J. R. 1940. Life history notes on the pilute ground squirrel. *J. Mammal.* **21**, 160–
170.

Allen, J. 1904. The association processes of the guinea pig. *J. Comp. Neurol. Psychol.* **14**,
293–359.

Altmann, M. 1952. Social behavior of elk, *Cervus canadensis,* in the Jackson Hole area of
Wyoming. *Behaviour* **4**, 116–143.

Altmann, M. 1956. Patterns of herd behavior in free-ranging elk of Wyoming *Cervus canadensis nelsoni. Zoologica (New York)* **41**, 65–71.

Altmann, M. 1960. The role of juvenile elk and moose in the social dynamics of their
species. *Zoologica (New York)* **45**, 35–39.

Altmann, M. 1961. Sex dynamics within kinships of free-ranging wild ungulates. Paper
read at Amer. Ass. Advan. Sci. Symp. on Incest, Denver.

Altmann, M. 1963. Naturalistic studies of maternal care in moose and elk. *In* "Maternal
Behavior in Mammals" (H. L. Rheingold, ed.), pp. 233–253. Wiley, New York.

Altmann, S. A. 1959. Field observations on a howling monkey society. *J. Mammal.* **40**, 317–
330.

Ambrose, J. A. 1961. The development of the smiling response in early infancy. *In* "Determinants of Infant Behaviour" (B. M. Foss, ed.), Vol. 1, pp. 179–196. Methuen,
London.

Ambrose, J. A. 1963. The concept of a critical period for the development of social re-
responsiveness in early human infancy. *In* "Determinants of Infant Behaviour"
(B. M. Foss, ed.), Vol. 2, pp. 201–225. Methuen, London.

Anglis, C. G. 1964. Breeding elephants *Elephas maximus* at the Budapest Zoo. *Int. Zoo Yearb.* **4**, 83–86.

Anonymous 1956. Behind the scenes. *Anim. Kingdom* **59**, 63.

Badham, M. 1967a. A note on the breeding of the spectacled leaf monkey. *Int. Zoo Yearb.* **7**, 89.

Badham, M. 1967b. A note on the breeding of the pileated gibbon *Hylobates lar pileatus* at Twycross Zoo. *Int. Zoo Yearb.* **7**, 92–93.

Bailey, V. 1924. Breeding, feeding and other life habits of meadow mice *(Microtus).* *J. Agr. Res.* **27**, 523–535.

Barbour, R. W., and Gault, W. L. 1953. *Peromyscus maniculatus bairdii* in Kentucky. *J. Mammal.* **34**, 130.

Bartholomew, G. A. 1952. Reproductive and social behavior of the Northern elephant seal. *Univ. Calif., Berkeley, Publ. Zool.* **47**, 369–472.

Bartholomew, G. A., and Hoel, P. G. 1953. Reproductive behavior of the Alaska fur seal, *Callorhinus ursinus. J. Mammal.* **34**, 417–436.

Bateson, P. P. G. 1966. The characteristics and context of imprinting. *Biol. Rev. Cambridge Phil. Soc.* **41**, 177–220.

Bateson, P. P. G. 1969. Imprinting and the development of preferences. *In* "Stimulation in Early Infancy" (A. Ambrose, ed.), pp. 109–125. Academic Press, New York.

Beach, F. A., and Jaynes, J. 1956a. Studies of maternal retrieving in rats. I. Recognition of young. *J. Mammal.* **37**, 177–180.

Beach, F. A., and Jaynes, J. 1956b. Studies of maternal retrieving in rats. II. Effects of practice and previous parturitions. *Amer. Natur.* **90**, 103–109.

Beach, F. A., and Jaynes J. 1956c. Studies of maternal retrieving in rats. III. Sensory cues involved in the lactating female's response to her young. *Behaviour* **10**, 104–125.

Beck, A. J., and Rudd, R. L. 1960. Nursery colonies in the pallid bat. *J. Mammal.* **41**, 266–267.

Bell, R. Q. 1964a. The problem of direction of effects in studies of parents and children. Conference on Research in Methodology in Parent-Child Interaction, under the auspices of the Dept. of Pediatrics, Upstate Medical Center, Syracuse, New York.

Bell, R. Q. 1964b. The effect on the family of a limitation in coping ability in a child: A Research approach and a finding. *Merrill-Palmer Quart.* **10**, 129–142.

Beniest-Noirot, E. 1958. Analyse du comportement dit maternal chez la souris. *Monogr. Fr. Psychol.* **1**.

Beniest-Noirot, E. 1962. Les modifications de réactivite au niveau des résponses aux jeunes chez la souris. Unpublished dissertation, Univ. Libre de Bruxelles, Belgium.

Bergman, S. 1936. Observations on the Kamchatkan bear. *J. Mammal.* **17**, 115–120.

Berlioz, J. 1933. L'evelage et l'hybridation du Bison au Canada. *Bull. Soc. Acclim. Fr.* **80**, 47–53.

Bernstein, T. S. 1964. A comparison of New and Old World monkey social organizations and behavior. *Amer. J. Phys. Anthropol.* **22**, 233–237.

Birkenholz, D. E., and Wirtz, W. O., II. 1965. Laboratory observations on the vesper rat. *J. Mammal.* **46**, 181–189.

Bishop, S. C. 1923. Note on the nest and young of the small brown weasel. *J. Mammal.* **4**, 26–27.

Blair, W. F. 1941. Observations on the life history of *Bairomys taylori subater. J. Mammal.* **22**, 378–383.

Blair, W. F. 1951. Population structure, social behavior, and environmental relations in a natural population of the beach mouse *(Peromyscus polionotus leucocephalus). Contrib. Lab. Vertebr. Biol., Univ. Mich.* **48**, 1–45.

Blair, W. F. 1958. Effects of X-irradiation on a natural population of the deer mouse *(Peromyscus maniculatus). Ecology* **39**, 113–118.

Blair, W. F., and Howard, W. E. 1944. Experimental evidence of sexual isolation between three forms of mice of the cenospecies *Peromyscus maniculatus. Contr. Lab. Vertebr. Biol., Univ. Mich.* **26**, 1–19.

Bolwig, N. 1958. A study of the behaviour of the chacma baboon, *Papio ursinus. Behaviour* **14**, 136–163.

Bonnot, P. 1928. Report on the seals and sea lions of California. *Fish Bull. Calif.* **14.**

Booth, C. 1962. Some observations on behaviour of Cercopithecus monkeys. *Ann. N. Y. Acad. Sci.* **102**, 477–487.

Bourlière, F. 1955. "The Natural History of Mammals." Harrap, London.

Bowden, D., Winter, P., and Ploog, D. 1967. Pregnancy and delivery behaviour in the squirrel monkey *(Saimiri sciureus)* and other primates. *Folia Primatol.* **5**, 1–42.

Brosset, A. 1966. "La biologie des Chiropteres." Masson, Paris.

Brown, R. Z. 1953. Social behavior, reproduction, and population changes in the house mouse *(Mus musculus L.). Ecol. Monogr.* **23**, 217–240.

Brownlee, A. 1957. Higher nervous activity in domestic cattle. *Brit. Vet. J.* **113**, 407–416.

Buechner, H. K. 1950. Life history, ecology, and range use of the Pronghorn antelope in Trans-Pecos Texas. *Amer. Midl. Natur.* **43**, 257–354.

Bullerman, R. 1950. Note on kidnapping by bonnet monkey. *J. Mammal.* **31**, 93–94.

Burckhardt, D. 1958. Observations sur la vie sociale du cerf *(Cervus elephas)* au parc National Suisse. *Mammalia* **22**, 226–244.

Cahalane, V. H. 1947. "Mammals of North America. Macmillan, New York.

Cairns, R. B. 1966. Attachment behavior of mammals. *Psychol. Rev.* **73**, 409–426.

Carpenter, C. R. 1934. A field study of the behavior and social relations of howling monkeys, *Alouatta palliata. Comp. Psychol. Monogr.* **10**, 1–168.

Carpenter, C. R. 1940. A field study in Siam of the behavior and social relations of the gibbon *(Hylobates lar). Comp. Psychol. Monogr.* **16**, 1–212.

Carpenter, C. R. 1965. The howlers of Barro Colorado Island. *In* "Primate Behavior" (I. DeVore, ed.), pp. 250–291. Holt, Rinehart & Winston, New York.

Carrick, R., Ingham, S. E., Csordas, S. E., and Keith, K. 1962. Studies on the South elephant seal, *Mirounga leonina. CSIRO Wildl. Res.* **7**, 89–206.

Castaret, N. 1938. Observations sur une colonie de chauves-souris migratrices. *Mammalia* **3**, 1–9.

Castaret, N. 1939. La colonie de murins de la grotte des Tignahustes. *Mammalia* **3**, 1–9.

Causey, D., and Waters, R. H. 1936. Parental care in mammals with special reference to the carrying of young by the albino rat. *J. Comp. Psychol.* **22**, 241–254.

Chaffee, P. S. 1967. A note on the breeding of orang-utans *Pongo pygmaeus* at Fresno Zoo. *Int. Zoo Yearb.* **7**, 94–95.

Clark, F. H. 1937. Parturition in the deer mouse. *J. Mammal.* **18**, 85–87.

Clark, M. J. 1968. Growth of pouch young in the red kangaroo in the pouches of foster mothers of the same species. *Int. Zoo Yearb.* **8**, 102–106.

Collias, N. E. 1956. The analysis of socialization in sheep and goats. *Ecology* **37**, 228–239.

Conoway, C. H. 1958. Maintenance, reproduction and growth of the least shrew in captivity. *J. Mammal.* **39**, 507–512.

Cooper, J. B. 1942. An exploratory study on African lions. *Comp. Psychol. Monogr.* **17**, 1–48.

Cosnier, J. 1963. Quelques problèmes posés par le "comportement maternal provoqué" chez la ratte. *C. R. Soc. Biol.* **157**, 1611–1613.

Cox, F. E. G. 1965. *Apodemus sylvaticus* fostered on *Mus musculus*. *Proc. Zool. Soc. London* **144**, 150–152.

Crandall, L. S. 1964. "Management of Wild Animals in Captivity." Univ. of Chicago Press, Chicago, Illinois.

Cross, H. A., and Harlow, H. F. 1963. Observations of infant monkeys by female monkeys. *Percept. Mot. Skills* **16**, 11–15.

Cuneo, F. 1965. Observations on the breeding of the Klipspringer antelope, *Oreotragus oreotragus*, and the behaviour of their young born at the Naples zoo. *Int. Zoo Yearb.* **5**, 45–48.

Darling, F. F. 1937. "A Herd of Red Deer." Oxford Univ. Press, Oxford.

Darling, F. F. 1938. "Wild Country." Cambridge Univ. Press, Cambridge, England.

Dathe, H. 1966. Breeding in the Corsac fox *Vulpes corsac* at East Berlin Zoo. *Int. Zoo Yearb.* **6**, 166–167.

David, R. 1966. Breeding the Indian wild ass *Equus hemionus khur* at Ahmedabad Zoo. *Int. Zoo Yearb.* **6**, 197–198.

Davis, J. A. 1965. A preliminary report of the reproductive behaviour of the small Malayan Chevrotain, *Tragulus javanicus* at New York Zoo. *Int. Zoo Yearb.* **5**, 42–44.

Davis, R. B., Herreid, C. F., II, and Short, H. L. 1962. Mexican free-tailed bats in Texas. *Ecol. Monogr.* **32**, 311–346.

Davis, W. H., Barbour, R. W., and Hassell, M. D. 1968. Colonial behavior of *Eptesicus fuscus. J. Mammal.* **49**, 44–50.

Deag, J. M. 1969. Personal communication.

Deansley, R., and Warwick, T. 1939. Observations on pregnancy in the common bat (*Pipistrellus pipistrellus*). *Proc. Zool. Soc. London* **109A**, 57–60.

de Kock, L. L. 1966. Breeding lemmings *Lemmus lemmus* for exhibition. *Int. Zoo Yearb.* **6**, 164–165.

Denenberg, V. H. 1962. The effects of early experience. *In* "The Behaviour of Domestic Animals" (E. S. E. Hafez, ed.), pp. 109–138. Baillière, London.

Denenberg, V. H., Sawin, P. B., Frommer, G. P., and Ross, S. 1958. Genetic, physiological and behavioral background of reproduction in the rabbit: IV. An analysis of maternal behavior at successive parturitions. *Behaviour* **13**, 131–141.

Denenberg, V. H., Grota, L. J., and Zarrow, M. X. 1962. Maternal behaviour in the rat: analysis of cross-fostering. *J. Reprod. Fert.* **5**, 133–141.

Denenberg, V. H., Hudgens, G. A., and Zarrow, M. X. 1966. Mice reared with rats: effects of mother on adult behavior patterns. *Psychol. Rep.* **18**, 451–456.

DeVore, I. 1963. Mother-infant relations in free-ranging baboons. *In* "Maternal Behavior in Mammals" (H. L. Rheingold, ed.), pp. 305–335. Wiley, New York.

DeVore, I., ed. 1965. "Primate Behavior." Holt, Rinehart & Winston, New York.

Dieterlin, F. 1959. Das verhalten des Syrischen Goldhamsters. *Z. Tierpsychol.* **16**, 47–103.

Dieterlin, F. 1960. Bermerkungen zu Zucht und verhalten der Zwergmaus (*Micromys minutus soricinus* Herman). *Z. Tierpsychol.* **17**, 552–554.

Dieterlin, F. 1962. Geburt und Geburtshiffe bie der Stachelmaus *Acomys cahirinus. Z. Tierpsychol.* **19**, 191–222.

Ditmars, R. L. 1933. Development of the silky marmoset. *Bull. N. Y. Zool. Soc.* **36**, 175–176.

Dittrich, L. 1968. Keeping and breeding gazelles at Hanover Zoo. *Int. Zoo Yearb.* **8**, 139–143.

Dobie, J. F. 1950. "The Voice of the Coyote." Hammond, Hammond, London.

Donald, H. P. 1937. Suckling and suckling preference in pigs. *Emp. J. Exp. Agr.* **5**, 361–368.

Douglas-Hamilton, I. 1969. Personal communication.

Doyle, G. A., Pelletier, A., and Bekker, J. 1967. Courtship, mating and parturition in the lesser bush baby (Galago senegalensis moholi) under semi-natural conditions. Folia Primatol. 7, 169–197.

Egoscue, H. J. 1956. Preliminary studies of the kit fox in Utah. J. Mammal. 37, 351–357.

Egoscue, H. J. 1957. The desert woodrat: A laboratory colony. J. Mammal. 38, 472–481.

Egoscue, H. J. 1964. Ecological notes and laboratory life history of the canyon mouse. J. Mammal. 45, 387–396.

Eibl-Eibesfeldt, I. 1950. Beiträge zur biologie der Haus-und Ährenmaus nebst einigen Beobachtungen an anderen Nagern. Z. Tierpsychol. 7, 558–587.

Eibl-Eibesfeldt, I. 1953. Zur Ethologie des Hamsters (Cricetus cricetus L.) Z. Tierpsychol. 10, 204–254.

Eibl-Eibesfeldt, I. 1955. Ethologische Studien am Galapagos-Seelöwen, Zalophus wollebacki Sivertson. Z. Tierpsychol. 12, 286–303.

Eisenberg, J. F. 1962. Studies on the behaviour of Peromyscus maniculatus gambelii and Peromyscus californicus parasiticus. Behaviour 19, 177–207.

Eisenberg, J. F. 1963. The intraspecific social behavior of some Cricetine rodents of the genus Peromyscus. Amer. Midl. Natur. 69, 240–246.

Eisentraut, M. 1936. Zur Fortpflanzungsbiologie der Fledermaüse. Z. Morphol. Oekol. Tiere 31, 27–63.

Epple, G. 1967. Vergleichende Untersuchungen über Sexual-und Sozialverhalten der Krallenaffin (Hapalidae). Folia Primatol. 7, 37–65.

Erickson, A. W., and Miller, L. H. 1963. Cub adoption in the brown bear. J. Mammal. 44, 384–385.

Espmark, Y. 1964a. Rutting behaviour in reindeer (Rangifer tarandus L.) Anim. Behav. 12, 159–163.

Espmark, Y. 1964b. Studies in dominance-subordination relationship in a group of semi-domestic reindeer (Rangifer tarandus L.). Anim. Behav. 12, 420–426.

Espmark, Y. 1968. Personal communication to W. H. Thorpe.

Essapian, F. S. 1963. Observations on abnormalities of parturition in captive bottle-nosed dolphins, Tursiops truncatus, and concurrent behavior of other porpoises. J. Mammal. 44, 405–414.

Estes, R. D. 1963. "Antelope Behavior Study," 3rd Rep., Aug. 1963–Jan. 1964. Nat. Geogr. Soc., Cornell Univ., Ithaca, New York.

Estes, R. D., and Goddard, J. 1967. Prey selection and hunting behavior of the African wild dog. J. Wildl. Manage. 31, 52–70.

Ewer, R. F. 1959. Suckling behaviour in kittens. Behaviour 15, 150–162.

Ewer, R. F. 1961. Further observations on suckling behaviour in kittens, together with some general considerations of the interrelations of innate and acquired responses. Behaviour 17, 247–260.

Ewer, R. F. 1963. The behaviour of the Meerkat, Suricata suricatta (Schreber). Z. Tier-psychol. 20, 570–607.

Ewer, R. F. 1967. The behaviour of the African Giant Rat (Cricetomys gambianus Water-house). Z. Tierpsychol. 24, 6–79.

Fisher, E. M. 1940. Early life of a sea-otter pup. J. Mammal. 21, 132–137.

Fitzgerald, A. 1935. Rearing marmosets in captivity. J. Mammal. 16, 181–188.

Fogden, S. C. L. 1968. Suckling behaviour in the grey seal (Halichoerus grypus) and the Northern elephant seal (Mirounga angustirostris). J. Zool. 154, 415–420.

Frank, F. 1952. Adoptionversuche bei Feld-mausen (Microtus arvalis Pall.). Z. Tierpsychol. 9, 415–423.

Fraser, A. 1937. "Sheep Farming." Crosby Lockwood, London.

Fuller, J. L., and DuBuis, E. M. 1962. The behaviour of dogs. In "The Behaviour of Domestic Animals" (E. S. E. Hafez, ed.), pp. 415–452. Ballière, London.

Gardner, B. T., and Wallach, L. 1965. Shapes of figures identified as a baby's head. Percept. Mot. Skills 20, 135–142.

Gashwiler, J. S., Robinette, W. L., and Morris, O. W. 1961. Breeding habits of bobcats in Utah. J. Mammal. 42, 76–84.

Gates, W. H. 1941. A few notes on the evening bat, Nycticeius humeralis (Rafinesque). J. Mammal. 22, 53–56.

Gauthier-Pilters, von H. 1959. Einige Beobachtungen zum Droh-, Angriffs- und Kampfverhalten des Dromedarhengstes, sowie über Geburt und nordwestlichen Sahara. Z. Tierpsychol. 16, 591–604.

Geist, V. 1963. On the behaviour of the North American moose (Alces alces andersoni Peterson 1950) in British Columbia. Behaviour 20, 377–416.

Gensch, W. 1962. Successful rearing of the binturong Arctitis binturong Raffl. Int. Zoo. Yearb. 4, 79–80.

Gillman, J. 1939. Some facts on captive Chacma baboons. J. Mammal. 20, 178–181.

Ginsberg, B. E. 1965. Coaction of genetical and nongenetical factors influencing sexual behavior. In "Sex and Behavior" (F. A. Beach, ed.), pp. 53–73. Wiley, New York.

Golley, F. B. 1957. Gestation period, breeding and fawning behavior of Columbian black-tailed deer. J. Mammal. 38, 116–120.

Goodpaster, W. W., and Hoffmeister, D. F. 1954. Life history of the golden mouse, Peromyscus nuttalli, in Kentucky. J. Mammal. 35, 16–27.

Gould, E., and Eisenberg, J. F. 1966. Notes on the biology of the Tenrecidae. J. Mammal. 47, 660–686.

Grange, W. B. 1932. Observations on the snowshoe hare, Lepus americanus phaeonotus Allen. J. Mammal. 13, 1–19.

Grinnell, J., Dixon, J. S., and Longdale, J. M. 1937. "Fur Bearing Mammals of California," Vols. I and II. Univ. of California Press, Berkeley, California.

Gucwinskand, H., and Gucwinski, A. 1968. Breeding the Zanzibar Galago at Kroclaw Zoo. Int. Zoo Yearb. 8, 111–114.

Hafez, E. S. E., and Schein, M. W. 1962. The behaviour of cattle. In "The Behaviour of Domestic Animals" (E. S. E. Hafez, ed.), pp. 247–296. Ballière, London.

Hafez, E. S. E., and Scott, J. P. 1962. The behaviour of sheep and goats. In "The Behaviour of Domestic Animals" (E. S. E. Hafez, ed.), pp. 297–333. Ballière, London.

Hafez, E. S. E., Sumption, L. J., and Jakway, J. S. 1962. The behaviour of swine. In "The Behaviour of Domestic Animals" (E. S. E. Hafez, ed.), pp. 334–369. Ballière, London.

Hahn, H. 1934. Die Familie der Procaviidae. Z. Saugetierk. 9, 207–358.

Hall, K. R. L., and DeVore, I. 1965. Baboon social behavior. In "Primate Behavior" (I. DeVore, ed.), pp. 53–110. Holt, Rinehart & Winston, New York.

Hall, K. R. L., and Mayer, B. 1967. Social interactions in a group of captive patas monkeys (Erythrocebus patas). Folia Primatol. 5, 213–236.

Hamilton, J. E. 1934. The Southern sea lion, Otaria byronica (De Blainville). Discovery Rep. 8, 269–318.

Hamilton, W. R. 1933. The weasels of New York. Amer. Midl. Natur. 14, 289–344.

Hampton, J. K. 1964. Laboratory requirements and observations of Oedipomidus oedipus. Amer. J. Phys. Anthropol. 22, 239–244.

Hampton, J. K., Hampton, S. H., and Landwehr, B. J. 1966. Observations on a successful breeding colony of the marmoset Oedipomidas oedipus. Folia Primatol. 4, 265–287.

Hanif, M. 1967. Notes on the breeding of the white-headed saki monkey (*Pithecia pithecia*) at Georgetown Zoo. *Int. Zoo Yearb.* **7**, 81–82.

Hansen, E. W. 1966. The development of maternal and infant behaviour in the rhesus monkey. *Behaviour* **27**, 107–149.

Hansen, R. M. 1957. Communal litters of *Peromyscus maniculatus*. *J. Mammal.* **38**, 523.

Hansen, W. R. 1958. Failure of mule deer to wean offspring. *J. Mammal.* **39**, 155.

Harlow, H. F., Harlow, M. K., and Hansen, E. W. 1963. The maternal affectional system of rhesus monkeys. *In* "Maternal Behavior in Mammals" (H. L. Rheingold, ed.), pp. 254–281. Wiley, New York.

Harper, J. A., Harn, J. H., Bentley, W. W., and Yocom, C. F. 1967. The status and ecology of the Roosevelt elk in California. *Wildl. Monogr.* **16**, 49 pp.

Hediger, H. 1958. Verhalten der Beuteltiere (Marsupialia). *Handb. Zool.* **8**, 10(9), 1–28.

Herscher, L., Moore, A. U., and Richmond, J. B. 1958. Effect of post-partum separation of mother and kid on maternal care in the domestic goat. *Science* **128**, 1342–1343.

Herscher, L., Richmond, J. B., and Moore, A. U. 1963. Maternal behavior in sheep and goats. *In* "Maternal Behavior in Mammals" (H. L. Rheingold, ed.), pp. 203–232. Wiley, New York.

Herter, K., and Herter, M. 1955. Über eine scheintrachtige Iltisfähe mit untergeschobenen katzenjungen. *Zool. Gart., Leipzig* **22**, 33–46.

Hess, E. H. 1959. Imprinting. *Science* **130**, 133–141.

Hill, W. C. O. 1957. "Primates III. Comparative Anatomy and Taxonomy, *Pithecoidea, Platyrrhini*." Edinburgh Univ. Press, Edinburgh.

Hill, W. C. O. 1966. Laboratory breeding, behavioural development and relations of the talapoin (*Miopithecus talapoin*). *Mammalia* **30**, 353–370.

Hinde, R. A. 1966. "Animal Behaviour." McGraw-Hill, New York.

Hinde, R. A., and Spencer-Booth, Y. 1967. The behaviour of socially living rhesus monkeys in their first two and a half years. *Anim. Behav.* **15**, 169–196.

Hinde, R. A., Rowell, T. E., and Spencer-Booth, Y. 1964. Behaviour of socially living rhesus monkeys in their first six months. *Proc. Zool. Soc. London* **143**, 609–649.

Hodge, W. H. 1946. Camels of the clouds. *Nat. Geogr. Mag.* **89**, 641–656.

Hopf, S. 1967. Ontogeny of social behaviour in the squirrel monkey. *1st Congr. Primatol. Soc. Frankfurt, 1966* pp. 255–262.

Horner, B. E. 1947. Paternal care of young mice of the genus *Peromyscus*. *J. Mammal.* **28**, 31–36.

Howard, W. E. 1949. Dispersal, amount of inbreeding, and longevity in a local population of prairie deermice on the George Reserve, Southern Michigan. *Contrib. Lab. Vertebr. Biol., Univ. Mich.* **43**, 1–50.

Itani, J. 1959. Paternal care in the wild Japanese monkey *Macaca fuscata*. *J. Primatol.*, **2**, 61–93.

Itani, J. 1961. The society of Japanese monkeys. *Jap. Quart.* **8**.

Jay, P. C. 1962. Aspects of maternal behavior among langurs. *Ann. N. Y. Acad. Sci.* **102**, 468–476.

Jay. P. C. 1963. Mother-infant relations in langurs. *In* "Maternal Behavior in Mammals" (H. L. Rheingold, ed.), pp. 282–304. Wiley, New York.

Jay, P. C. 1965. The common langur of North India. *In* "Primate Behavior" (I. DeVore, ed.), pp. 197–249. Holt, Rinehart & Winston, New York

Kahmann, H., and Frisch, O. von 1952. Über die Beziehungen von Muttertier und Nestling bei kleinen Saügetiern. *Experientia* **8**, 221–223.

Katz, I. 1949. Behavioral interactions in a herd of Barbary sheep. *Zoologica (New York)* **34**, 9–18.

Kaufmann, J. H. 1965. Studies on the behaviour of captive tree shrews *(Tupaia glis). Folia Primatol.* **3**, 50–74.

Kaufmann, J. H. 1966. Behavior of infant rhesus monkeys and their mothers in a free-ranging band. *Zoologica (New York)* **51**, 17–27.

Kaye, S. V. 1961. Laboratory life history of the eastern harvest mouse. *Amer. Midl. Natur.* **66**, 439–451.

King, J. A. 1955. Social behavior, social organization, and population dynamics in a black-tailed prairie dog town in the Black Hills of South Dakota. *Contrib. Lab. Vertebr. Biol., Univ. Mich.* **67**, 1–123.

King, J. A. 1956. Social relations of the domestic guinea pig living under semi-natural conditions. *Ecology* **37**, 221–228.

King, J. A. 1958. Maternal behavior and behavioral development in two sub-species of *Peromyscus maniculatus. J. Mammal.* **39**, 177–190.

King, J. A. 1963. Maternal behavior in *Peromyscus. In* "Maternal Behavior in Mammals" (H. L. Rheingold, ed.), pp. 58–93. Wiley, New York.

Kleiman, D. G. 1968. Reproduction in the Canidae. *Int. Zoo Yearb.* **8**, 3–8.

Kleiman, D. G. 1969a. The reproductive behaviour of the green acouchi, *Myoprocta pratti.* Unpublished dissertation, Univ. of London.

Kleiman, D. G. 1969b. Maternal care, growth rate, and development in the noctule *(Nyctalus noctula),* pipistrelle *(Pipistrellus pipistrellus),* and serotine *(Eptesicus serotinus)* bats. *J. Zool.* **157**, 187–211.

Kleiman, D. G. 1970. Personal communication.

Klingel, von H., and Klingel, V. 1966. Die Geburt eines Zebras *(Equus quagga böhmi* Matschi). *Z. Tierpsychol.* **23**, 72–76.

Klopfer, P. H., and Gamble, J. 1966. Maternal "imprinting" in goats: the role of chemical senses. *Z. Tierpsychol.* **23**, 588–592.

Klopfer, P. H., and Gilbert, B. K. 1966. A note on retrieval and recognition of the young in the elephant seal, *Mirounga angustirostris. Z. Tierpsychol.* **23**, 757–760.

Klopfer, P. J., Adams, D. K., and Klopfer, M. S. 1964. Maternal "imprinting" in goats. *Proc. Nat. Acad. Sci. U.S.* **52**, 911–914.

Koford, C. B. 1957. The vicuña and the puna. *Ecol. Monogr.* **27**, 153–219.

Kühme, von W. 1965a. Freilandstudien zur Soziologie des Hyaenenhundes *(Lycaon pictus lupinus* Thomas 1902). *Z. Tierpsychol.* **22**, 1–541.

Kühme, von W. 1965b. Communal food distribution and division of labour in African hunting dogs. *Nature (London)* **205**, 443–444.

Kulzer, E. 1958. Untersuchungen über die Biologie von Flughunden der Gattung *Roussetus* Gray. *Z. Morphol. Oekol. Tiere* **47**, 374–402.

Kummer, H. 1967. Tripartite relations in Hamadryas baboons. *In* "Social Communication among Primates" (S. A. Altmann, ed.), pp. 63–71. Univ. of Chicago Press, Chicago, Illinois.

Kummer, H. 1968. "Social Organization of Hamadryas Baboons." Univ. of Chicago Press, Chicago, Illinois.

Kunkel, von P., and Kunkel, I. 1964. Beitrage zur Ethologie des Haus meerschweinchens *Cavia aperea f. porcellus* (L.). *Z. Tierpsychol.* **21**, 602–641.

Lamond, H. G. 1949. Mothering a lamb. *Sheep Goat Raising* **29**, 36–38.

Langford, J. B. 1963. Breeding behaviour of *Hapale jacchus* (common marmoset). *S. Afr. J. Sci.* **59**, 299–230.

Laws, R. M. 1956. The elephant seal (Mirounga leonina Lina.). II. General, social and reproductive behaviour. Falkland Isl. Depend. Surv., Sci. Rep. 13, 88 pp.

Laws, R. M., and Clough, G. 1966. Observations on reproduction in the hippopotamus Hippopotamus amphibius Linn. In "Comparative Biology of Reproduction in Mammals" (I. W. Rowlands, ed.), Symp. Zool. Soc. London No. 15, pp. 117–140. Academic Press, New York.

Layne, J. N. 1959. Growth and development of the Eastern harvest mouse Reithrodontomys humulis. Bull. Fla. State Mus., Biol. Sci. 4, 61–82.

Leblond, C. P. 1938. Extra-hormonal factors in maternal behavior. Proc. Soc. Exp. Biol. Med. 38, 66–70.

Leblond, C. P. 1940. Nervous and hormonal factors in the maternal behavior of the mouse. J. Genet. Psychol. 57, 327–344.

Lehrman, D. S. 1961. Gonadal hormones and parental behavior in birds and infrahuman mammals. In "Sex and Internal Secretions" (W. C. Young, ed.), pp. 1268–1382. Williams & Wilkins, Baltimore, Maryland.

Lent, P. C. 1966. Calving and related social behaviour in the barren ground caribou. Z. Tierpsychol. 23, 701–756.

Liers, E. E. 1951a. Notes on the river otter (Lutra canadensis). J. Mammal. 32, 1–9.

Liers, E. E. 1951b. My friends the land otters. Natur. Hist. 60, 320–326.

Linsdale, J. M., and Tomich, P. Q. 1953. "A Herd of Mule Deer." Univ. of Calif. Press, Berkeley, California.

Lloyd, J. I. 1951. When do otters breed? Field 197, 1035.

Lockley, R. M. 1966. "Grey Seal, Common Seal." Deutsch, London.

Loisel, G. 1906. Relations entre les phénomènes du rut, de la lactation, de la lune et de l'amour maternal chez une chienre hybride. C. R. Soc. Biol. 60, 255–258.

Lott, D. F., and Fuchs, S. S. 1962. Failure to induce retrieving by sensitization or the injection of prolactin. J. Comp. Physiol. Psychol. 55, 1111–1113.

Lowther, F. de L. 1940. A study of the activities of a pair of Galago senegalensis moholi in captivity including the birth and post-natal development of twins. Zoologica. (New York) 25, 433–462.

Lucas, N. S., Hume, E. M., and Henderson Smith, H. 1927. On the breeding of the common marmoset (Hapale jacchus Linn.) in captivity when irradiated with ultra-violet rays. Proc. Zool. Soc. London 30, 447–451.

Lucas, N. S., Hume, E. M., and Henderson Smith, H. 1937. On the breeding of the common marmoset (Hapale jacchus Linn.) in captivity when irradiated with ultra-violet rays. II. A ten years' family history. Proc. Zool. Soc. London, Ser. A 107, 205–211.

Maberley, C. T. A. 1966. "Animals of East Africa." Hodder & Stoughton, London.

Maberry, M. B. 1964. Breeding Indian elephants, Elephas maximus at Portland Zoo. Int. Zoo Yearb. 4, 81–83.

McBride, A. F., and Hebb, D. O. 1948. Behavior of the captive bottle-nose dolphin Tursiops truncatus. J. Comp. Physiol. Psychol. 41, 111–123.

McBride, A. F., and Kritzler, H. 1951. Observations on pregnancy, parturition and post-natal behavior in the bottle-nose dolphin. J. Mammal. 32, 251–266.

McBride, G. 1963. The "teat order" and communication in young pigs. Anim. Behav. 11, 53–56.

McCann, C. 1934. Observations on some of the Indian langurs. J. Bombay Natur. Hist. Soc. 36, 616–628.

McCann, L. J. 1956. Ecology of the mountain sheep. Amer. Midl. Natur. 56, 297–324.

McHugh, T. 1958. Social behavior of the American buffalo (Bison bison bison). Zoologica (New York) 43, 1–40.

Martin, R. D. 1966. Tree shrews: unique reproductive mechanism of systematic importance. *Science* **152**, 1402–1404.

Mathison, O. A., Beade, R. T., and Lopp, R. J. 1962. Breeding habits, growth and stomach
contents of the Stellar sea lion in Alaska. *J. Mammal.* **43**, 469–477.

Matthews, L. H. 1937. The female sexual cycle in the British horse-shoe bats. *Trans. Zool.
Soc. London* **23**, 224–266.

Mendelssohn, H. 1965. Breeding Syrian hyrax *Procavia capensis syriaca* Schreber 1784. *Int.
Zoo Yearb.* **5**, 116–125.

Menzel, R., and Menzel, R. 1953. Einiges aus der Pflegewelt der Mutterhündin. *Behaviour*
5, 289–304.

Merchant, J. C., and Sharman, G. B. 1966. Observations on the attachment of marsupial
pouch young by foster mothers of the same or different species. *Aust. J. Zool.* **14**,
593–609.

Meyer, B. J., and Meyer, R. K. 1944. Growth and reproduction of the cotton rat, *Sigmodon
hispidus hispidus*, under laboratory conditions. *J. Mammal.* **25**, 107–129.

Mohr, C. E. 1933. Observations on the young of cave-dwelling bats. *J. Mammal.* **14**, 49–53.

Moltz, H., and Wiener, E. 1966. Effects of ovariectomy on maternal behavior of primiparous and multiparous rats. *J. Comp. Physiol. Psychol.* **62**, 382–387.

Moltz, H., Robbins, D., and Parks, M. 1966. Caesarian delivery and maternal behavior of
primiparous and multiparous rats. *J. Comp. Physiol. Psychol.* **61**, 455–460.

Moore, A. U., and Moore, F. 1960. Studies in the formation of the mother-neonate bond in
sheep and goats. Symposium on Mechanisms of Primary Socialization. *Amer. Psychol.*
15, 413.

Morris, D. 1965. "The Mammals." Hodder & Stoughton, London.

Moynihan, M. 1964. Some behavior patterns of Platyrrhine monkeys. I. The night monkey
(Aotus trivirgatus). Smithson. Misc. Collect. **146**, No. 5.

Murie, A. 1944. The wolves of Mt. McKinley. *Nat. Park Serv., U.S. Dept. Interior Fauna Ser.* **5.**

Napier, J. R., and Napier, P. N. 1967. "A Handbook of Living Primates." Academic, N. Y.

Neal, B. J. 1959. A contribution to the life history of the collared peccary in Arizona.
Amer. Midl. Natur. **61**, 177–190.

Nelson, W. W., and Smelser, G. K. 1933. Studies on the physiology of lactation. II. Lacta-
in the male guinea pig and its bearing on the corpus luteum problem. *Amer. J. Physiol.*
103, 374–381.

Nevo, E., and Amir, E. 1964. Geographic variation in reproduction and hibernation patterns of the forest dormouse. *J. Mammal.* **45**, 69–87.

Nicholson, A. J. 1941. The homes and social habits of the wood mouse *(Peromyscus leucopus noveboracensis). Amer. Midl. Natur.* **25**, 196–223.

Nissen, H. W. 1931. A field study of the chimpanzee. Observations of chimpanzee behavior
and environment in Western French Guinea. *Comp. Psychol. Monogr.* **8**, 1–105.

Noirot, E. 1964a. Changes in responsiveness to young in the adult mouse: The effect of
external stimuli. *J. Comp. Physiol. Psychol.* **57**, 97–99.

Noirot, E. 1964b. Changes in responsiveness to young in the adult mouse. IV. The effect
of an initial contact with a strong stimulus. *Anim. Behav.* **12**, 442–445.

Noirot, E. 1964c. Changes in responsiveness to young in the adult mouse. I. The problematical effect of hormones. *Anim. Behav.* **12**, 52–58.

Noirot, E. 1965. Changes in responsiveness to young in the adult mouse. III. The effect of
immediately preceding performances. *Behaviour* **14**, 318–325.

Noirot, E. 1968. Personal communication.

Noirot, E. 1969. Changes in responsiveness to young in the adult mouse. V. Priming.
Anim. Behav. **17**, 542–546.

Noirot, E., and Richards, M. P. M. 1966. Maternal behaviour in virgin female golden hamsters: changes consequent upon initial contact with pups. *Anim. Behav.* **14**, 7–10.

Orr, R. T. 1954. Natural history of the Pallid bat, *Antrozous pallidus* (Le Conte). *Proc. Calif. Acad. Sci.* **28**, 165–246.

Pearson, O. P. 1948. Life history of mountain viscachas in Peru. *J. Mammal.* **29**, 345–374.

Pearson, O. P., Koford, M. R., and Pearson, A. K. 1952. Reproduction in the lump-nosed bat *(Corynorhinus rafinesqui)* in California. *J. Mammal.* **33**, 273–320.

Pearson, P. G. 1952. Observations concerning the life history and ecology of the wood rat, *Neotoma floridana floridana* (Ord). *J. Mammal.* **33**, 459–463.

Petter, J. J. 1965. The lemurs of Madagascar. *In* "Primate Behavior" (I. DeVore, ed.), pp. 292–319. Holt, Rinehart & Winston, New York.

Pilters, H. 1954. Untersuchungen über angeborene verhaltensweisen bei Tylopoden, unter besonderer Berücksichtigung der neuweltlichen Formen. *Z. Tierpsychol.* **11**, 213–303.

Polderbeer, E. B., Kuhn, L. W., and Hendrickson, G. O. 1941. Winter and spring habits of weasels in central Iowa. *J. Wildl. Manage.* **5**, 115–119.

Poppleton, F. 1957. Birth of an elephant. *Oryx* **4**, 180–181.

Pournelle, G. H. 1967. Observation on reproductive behaviour and early post-natal development of the proboscis monkey *Nasalis larvatus orientalis* at San Diego Zoo. *Int. Zoo Yearb.* **7**, 90–92.

Preble, E. A. 1923. "Mammals of the Pribilof Islands, North American Fauna No. 46: A Biological Survey of the Pribilof Islands, Alaska. Birds and Mammals." U.S. Dept. Agric. Gov. Printing Office, Washington, D.C.

Prior, R. 1968. "The Roe Deer of Cranbourne Chase." Oxford Univ. Press, London and New York.

Pruitt, W. O. 1960. Behavior of the barren-ground caribou. *Biol. Pap. Univ. Alaska* **3**, 1–44.

Racey, P. 1969. Personal communication.

Ransom, T., and Ransom, B. 1969. Personal communication.

Ressler, R. H. 1962. Parental handling in two strains of mice reared by foster parents. *Science* **137**, 129–130.

Reynolds, V., and Reynolds, F. 1965. Chimpanzees in the Budongo forest. *In* "Primate Behavior" (I. DeVore, ed.), pp. 368–424. Holt, Rinehart & Winston, New York.

Rheingold, H. L. 1963. Maternal behavior in the dog. *In* "Maternal Behavior in Mammals" (H. L. Rheingold, ed.), pp. 169–202. Wiley, New York.

Rice, D. W. 1957. Life history and ecology of *Myotis austroriparius* in Florida. *J. Mammal.* **38**, 15–32.

Richards, M. P. M. 1966a. Maternal behaviour in the golden hamster: responsiveness to young in virgin, pregnant, and lactating females. *Anim. Behav.* **14**, 310–313.

Richards, M. P. M. 1966b. Maternal behaviour in virgin female golden hamsters *(Mesocricetus auratus* Waterhouse): the role of the age of the test pup. *Anim. Behav.* **14**, 303–309.

Richards, M. P. M. 1967. Maternal behaviour in rodents and lagomorphs. *Advan. Reprod. Physiol.* **2**, 53–110.

Riddle, O., Lahr, E. L., and Bates, R. W. 1942. The role of hormones in the initiation of maternal behavior in rats. *Amer. J. Physiol.* **137**, 299–317.

Rosenblatt, J. S. 1965. The basis of synchrony in the behavioral interaction between the mother and her offspring in the laboratory rat. *In* "Determinants of Infant Behaviour" (B. M. Foss, ed.), Vol. 3, pp. 3–45. Methuen, London.

Rosenblatt, J. S. 1967. Non-hormonal basis of maternal behavior. *Science* **156**, 1512–1514.

Rosenblatt, J. S., and Lehrman, D. S. 1963. Maternal behavior of the laboratory rat. *In* "Maternal Behavior in Mammals" (H. L. Rheingold, ed.), pp. 8–57. Wiley, New York.

Rosenblatt, J. S., and Schneirla, T. C. 1962. The behaviour of cats. *In* "The Behaviour of Domestic Animals" (E. S. E. Hafez, ed.), pp. 453–488. Baillière, London.

Rosenblatt, J. S., Turkewitz, G., and Schneirla, T. C. 1961. Early socialization in the domestic cat as based on feeding and other relationships between mother and young. *In* "Determinants of Infant Behaviour" (B. M. Foss, ed.), pp. 51–74. Methuen, London.

Ross, S., Denenberg, V. H., Frommer, G. P., and Sawin, P. B. 1959. Genetic, physiological and behavioral background of reproduction in the rabbit. V. Non-retrieving of neonates. *J. Mammal.* **40**, 91–96.

Roth, C. E. 1957. Notes on maternal care in *Myotis lucifugus. J. Mammal.* **38**, 122–123.

Roth-Kolar, H. 1957. Beiträge zu einem Aktionssystem des Aguti. *Z. Tierpsychol.* **14**, 362–375.

Rowell, T. E. 1960. On the retrieving of young and other behaviour in lactating golden hamsters. *Proc. Zool. Soc. London* **135**, 265–282.

Rowell, T. E. 1961a. Maternal behaviour in non-maternal golden hamsters *(Mesocricetus auratus). Anim. Behav.* **9**, 11–15.

Rowell, T. E. 1961b. The family group in golden hamsters: its formation and break up. *Behaviour* **17**, 81–94.

Rowell, T. E., Hinde, R. A., and Spencer-Booth, Y. 1964. "Aunt"-infant interaction in captive rhesus monkeys. *Anim. Behav.* **12**, 219–226.

Rowley, J. 1929. Life history of the sea-lions of the California coast. *J. Mammal.* **10**, 1–36.

Sanderson, I. T. 1957. "The Monkey Kingdom." Hamish-Hamilton, London.

Sauer, E. G. F. 1967. Mother-infant relationship in galagos and the oral child-transport among Primates. *Folia Primatol.* **7**, 127–149.

Schaffer, H. R. 1963. Some issues for research in the study of attachment behaviour. *In* "Determinants of Infant Behaviour" (B. M. Foss, ed.), pp. 179–196. Methuen, London.

Schaller, G. B. 1965. The behavior of the mountain gorilla. *In* "Primate Behavior" (I. DeVore, ed.), pp. 324–367. Holt, Rinehart & Winston, New York.

Schenkel, R. 1966. Play, exploration and territoriality in the wild lion. *In* "Play, Exploration and Territory in Mammals" (P. A. Jewell, ed.), Symp. Zool. Soc. London No. 18, pp. 11–22. Academic Press, New York.

Schneirla, T. C., and Rosenblatt, J. S. 1961. Behavioral organization and genesis of the social bond in insects and mammals. *Amer. J. Orthopsychiat.* **31**, 223–253.

Schneirla, T. C., Rosenblatt, J. S., and Tobach, E. 1963. Maternal behavior in the cat. *In* "Maternal Behavior in Mammals" (H. L. Rheingold, ed.), pp. 122–168. Wiley, New York.

Schoonmaker, W. J. 1938. The woodchuck: Lord of the clover field. *Bull. N.Y. Zool. Soc.* **41**, 3–12.

Scott, J. P. 1945. Social behavior, organization and leadership in a small flock of domestic sheep. *Comp. Psychol. Monogr.* **18**, 1–29.

Scott, J. P. 1950. The social behavior of dogs and wolves: an illustration of sociobiological systematics. *Ann. N.Y. Acad. Sci.* **51**, 1009–1021.

Scott, J. P. 1960. Comparative social psychology. *In* "Principles of Comparative Psychology" (R. H. Waters, D. A. Rethlinshafer, and W. E. Caldwell, eds.), pp. 250–288. McGraw-Hill, New York.

Scott, J. P. 1962. Introduction to animal behaviour. *In* "The Behaviour of Domestic Animals" (E. S. E. Hafez, ed.), pp. 3–20. Baillière, London.

Seton, E. T. 1910. "Life Histories of Northern Mammals," Vols. I and II. Constable, London.

Seton, E. T. 1929. "Lives of Game Animals," Vols. I and II. Hodder & Stoughton, London.

Severinghaus, C. W., and Cheatum, E. L. 1956. Life and times of the white-tailed deer. *In* "The Deer of North America" (W. P. Taylor, ed.), pp. 57–186. Stackpole, Harrisburg, Pennsylvania.

Seward, J. P., and Seward, G. H. 1940. Studies on the reproductive activities of the guinea pig. *J. Comp. Psychol.* **29**, 1–24.

Sharman, G. B. 1962. The initiation and maintenance of lactation in the marsupial *Trichosurus vulpecula*. *J. Endocrinol.* **25**, 373–385.

Sharman, G. B., and Calaby, J. H. 1964. Reproductive behaviour in the red kangaroo, *Megaleia rufa* in captivity. *CSRIO Wildl. Res.* **9**, 58–85.

Sheldon, W. G. 1949. Reproductive behavior of foxes in New York State. *J. Mammal.* **30**, 236–246.

Simonds, P. E. 1965. The bonnet macaque in South India. *In* "Primate Behavior" (I. DeVore, ed.), pp. 175–196. Holt, Rinehart & Winston, New York.

Smith, C. C. 1968. The adaptive nature of social organization in the genus of tree squirrels *Tamiasciureus*. *Ecol. Monogr.* **38**, 31–63.

Smith, E. A. 1968. Adoptive suckling in the grey seal. *Nature (London)* **217**, 762–763.

Sollberger, D. E. 1940. Notes on the life history of the small eastern flying squirrel. *J. Mammal.* **21**, 282–293.

Sorenson, M. W., and Conaway, C. H. 1966. Observations on the social behaviour of tree shrews in captivity. *Folia Primatol.* **4**, 124–145.

Southern, H. N., ed. 1964. "The Handbook of British Mammals." Blackwell, Oxford.

Southwick, C. H., Beg, M. A., and Siddiqui, M. R. 1965. Rhesus monkeys in North India. *In* "Primate Behavior" (I. DeVore, ed.), pp. 111–159. Holt, Rinehart & Winston, New York.

Spencer, C. C. 1943. Notes on the life history of rocky mountain bighorn sheep in the Tarryall mountains of Colorado. *J. Mammal.* **24**, 1–11.

Spencer-Booth, Y. 1968. The behaviour of group companions towards rhesus monkey infants. *Anim. Behav.* **16**, 541–557.

Spencer-Booth, Y., and Hinde, R. A. 1967. The effects of separating rhesus monkey infants from their mothers for six days. *J. Child Psychol. Psychiat.* **7**, 179–197.

Stebbings, R. E. 1966. A population study of the bats of the genus *Plecotus*. *J. Zool.* **150**, 53–75.

Stegeman, Le Roy L. 1937. Notes on young skunks in captivity. *J. Mammal.* **18**, 194–202.

Stegeman, Le Roy L. 1938. The European wild boar in the Cherokee national forest, Tennessee. *J. Mammal.* **19**, 279–290.

Stellar, E. 1960. The marmoset as a laboratory animal: maintenance, general observations of behavior, and simple learning. *J. Comp. Physiol. Psychol.* **53**, 1–10.

Stevens, M. N. 1957. "The Otter Report." U.F.A.W., London.

Stodart, E. 1966. Management and behaviour of breeding groups of the marsupial *Perameles nasuta* Geoffroy in captivity. *Aust. J. Zool.* **14**, 611–623.

Struhsaker, T. T. 1967a. Behavior of vervet monkeys *(Cercopithecus aethiops)*. *Univ. Calif., Berkeley, Publ. Zool.* **82**, 1–74.

Struhsaker, T. T. 1967b. Behavior of elk *(Cervus canadensis)* during the rut. *Z. Tierpsychol.* **24**, 80–114.

Sugiyama, Y. 1965a. On the social change of Hanuman langurs *(Presbytis entellus)* in their natural condition. *Primates* **6**, 382–418.

Sugiyama, Y. 1965b. Behavioural development and social structure in two troops of Hanuman langurs. *Primates* **6**, 213–217.

Sugiyama, Y. 1967. Social organization of Hanuman langurs. *In* "Social Communication among Primates" (S. A. Altmann, ed.), pp. 221–236. Univ. of Chicago Press, Chicago, Illinois.

Svihla, A. 1932. A comparative life history study of the mice of the genus *Peromyscus*. *Univ. Mich. Mus. Zool., Misc. Publ.* **24**, 1–39.

Talbot, L. M., and Talbot, M. H. 1963. The wildebeest in Western Masailand, East Africa. *Wildl. Monogr.* **12**, 1–88.

Tavolga, M. C., and Essapian, F. S. 1957. The behavior of the bottle-nosed dolphin *(Tursiops truncatus)*: mating, pregnancy, parturition and mother-infant behavior. *Zoologica (New York)* **42**, 11–31.

Tevis, L. 1950. Summer behavior of a family of beavers in New York State. *J. Mammal.* **31**, 40–65.

Thompson, J. A., and Owen, W. H. 1964. A field study of the Australian ringtail possum *Pseudocheirus peregrinus* (Marsupialia: Phalangeridae). *Ecol. Monogr.* **34**, 27–52.

Thorpe, W. H. 1968. Perceptual basis for group organisation in social vertebrates, especially birds. *Nature (London)* **220**, 124–128.

Tyler, S. 1969. Personal communication.

van Kirchshofer, R. 1960. Einige Verhaltensbeobachtungen an Einem Guereza-Jungen *Colobus polykomos kikuyuensis* unter besonderer Berücksichtigung des Spiels. *Z. Tierpsychol.* **17**, 506–514.

van Lawick, H. 1969. Personal communication.

van Lawick-Goodall, J. 1965. Infancy, childhood and adolescence in the wild chimpanzee. *Proc. Roy. Inst. Gt. Brit.* **40**, 518–529.

van Lawick-Goodall, J. 1967. Mother-offspring relations in free-ranging chimpanzees. *In* "Primate Ethology" (D. Morris, ed.), pp. 287–346. Weidenfeld & Nicolson, London.

van Lawick-Goodall, J. 1968. The behaviour of free-living chimpanzees in the Gombe Stream area. *Anim. Behav. Monogr.* **1**, 161–311.

von Haas, G. 1959. Untersuchungen über angeborene verhaltensweisen bei Mähnenspringern *(Ammotragus lervia Pallas)*. *Z. Tierpsychol.* **16**, 218–242.

von Koenig, L. 1960. Das Aktionssystem des Siebenschläfers *(Glis glis L.)*. *Z. Tierpsychol.* **17**, 425–505.

von Walther, F. 1964. Verhaltensstudien an der Gattung *Tragelaphus* de Blainville, 1816 in Gefangenschaft, unter besonderer Berücksichtigung des Sozialverhaltens. *Z. Tierpsychol.* **21**, 393–467.

von Zannier, F. 1965. Verhaltensuntersuchungen an der Zwermanguste *Helogale undulata rufula* in Zoologischen Garten Frankfurt am Main. *Z. Tierpsychol.* **22**, 672–695.

Wackernagel, H. 1965. Grant's zebra, *Equus burchelli boehmi*, at Basle Zoo—a contribution to breeding biology. *Int. Zoo Yearb.* **5**, 38–41.

Wagner, H. O. 1956. Freilandbeobachtungen an Klammeraffen. *Z. Tierpsychol.* **13**, 302–313.

Washburn, S. L., Jay, P. C., and Lancaster, J. B. 1965. Field studies of Old World monkeys and apes. *Science* **150**, 1541–1547.

Westell, W. P. 1910. "The Book of the Animal Kingdom." Dent, London.

Wiesner, B. P., and Sheard, N. M. 1933. "Maternal Behaviour in the Rat." Oliver & Boyd, Edinburgh and London.

Williams, L. 1967. Breeding Humboldt's woolly monkey at the Murrayton woolly monkey sanctuary. *Int. Zoo Yearb.* **7**, 86–89.

Wimsatt, W. A., and Guerriere, A. 1961. Care and maintenance of the common vampire in captivity. *J. Mammal.* **42**, 449–455.

Yadav, R. N. 1968. Notes on breeding the Indian wolf in Jaipur Zoo. *Int. Zoo Yearb.* **8,** 17–18.

Yardeni-Yaron, E. 1952. [*Microtus guentheri*]. *Bull. Res. Counc. Isr.* **1,** 96–98.

Young, S. P. 1958. "The Bobcat of North America." Stackpole, Harrisburg, Pennsylvania.

Young, S. P., and Goldman, E. A. 1944. "The Wolves of North America." Amer. Wildl. Inst., Washington, D. C.

Zoonooz. 1964. **37** No. 6, 7. [*Colobus colobus*].

Tool-Using in Primates and Other Vertebrates

JANE VAN LAWICK-GOODALL

GOMBE STREAM RESEARCH CENTRE
KIGOMA, TANZANIA

I. INTRODUCTION

Because tool-using has played a major role in the evolution of man, much attention has been directed to the use of objects as tools in lower animals. Anthropologists believe that a more detailed knowledge of tool use in primates, and of its origins, in ontogeny and phylogeny, will help to shed light on the development of tool use in early man. Behaviorists and psychologists have been primarily concerned with demonstrating either that tool-using behavior in nonhuman animals provides evidence of insight and "intelligence," or that it can be explained as purely "innate behavior." In fact, tool-using performances occur in species that are widely separated in the phylogenetic scale, and the evolutionary processes that have led to tool-using in, say, an insect on the one hand, and man on the other, are undoubtedly very different.

There is no general agreement in the literature as to what is, and what is not, tool-using in animals. In this chapter, a tool-using performance in an animal or bird is specified as the use of an external object as a functional extension of mouth or beak, hand or claw, in the attainment of an immediate goal. This goal may be related to the obtaining of food, care of the body, or repulsion of a predator, intruder, etc. If the object is used successfully, then the animal achieves a goal which, in a number of instances, would not have been possible without the aid of the tool.

It may be helpful, at the outset, to make a clear distinction between a "tool" and "material." Tailor birds (*Orthotomus* spp.) have sometimes been considered to be tool-using birds (e.g., Thorpe, 1956). These birds make their nests within the fold of a large hanging leaf, the edges of which are "sewn" together with plant fiber. However, although this is undoubtedly a skillful manipulation of materials, it does not differ in any fundamental way from similar manipulations shown by other birds during nest-building. Some weavers (Ploceidae), for instance, may knot strands of grass around twigs with half-hitches (Crook, 1960).

It has actually been suggested (Lancaster, 1968) that we should perhaps consider the nest-making activities of all birds and mammals as examples of tool-using behavior. However, although one might argue, with Lancaster, that the finished product is a "tool" for sleeping or for the raising of young, it seems more logical to regard the beak and claws or mouth and paws of the bird or mammal as the tools in such cases, and the straws and leaves as the materials being manipulated. When someone is knitting, for instance, it is the knitting needles and not the wool or the finished jersey that are normally regarded as the tools.

Another category of behavior that is sometimes considered as tool or implement using is the string-pulling behavior of *Parus* spp. and some other passerines (e.g., Thorpe, 1956), monkeys (Klüver, 1933), and apes (e.g., Köhler, 1925). This again should perhaps be considered as skillful manipulation of objects rather than true tool use. The string and the food lure form a visual continuum and, in pulling on the string, the animal is merely pulling at a part of the food. A budgerigar will pull in a millet head if the end of the stem is put within its reach (personal observation): primates, in the wild, may pull lengths of vine hand-over-hand in order to reach shoots on the end (e.g., chimpanzees and baboons, personal observation; mangabeys, *Cercocebus albigena,* Chalmers, personal communication); and all monkeys and apes will break off or bend over branches in order to feed on fruit at their peripheries. Similar behavior is shown by elephants when they push over trees to feed on foliage otherwise out of reach.

It may put tool-using behavior, as such, in a better perspective if, before reviewing our present-day knowledge of the behavior in primates we briefly consider what is known of tool-using performances in non-primate mammals and in birds.[1] In doing so I shall include brief refer-

[1] There are a few examples of tool-using in invertebrates, and one fish "shoots" down insects by spitting water at them from the surface of the water. Some of these are mentioned by Hall (1963). But the evolutionary gap between an insect and a primate is too great to make it useful to review such performances in this chapter.

ences to types of behavior that seem to me to bear a close resemblance to tool-using, though not actually falling within my definition.

II. TOOL-USING AND RELATED BEHAVIOR IN BIRDS

Several species of birds are known to use tools regularly, or fairly frequently, in order to obtain food, and one species shows this behavior during courtship. The other examples given below refer to instances of tool-use observed once or twice in a few individuals only, but in many cases further research may show that the behavior occurs more generally within the species. A variety of birds show behavior closely related to tool-using.

a. Use of Thorn or Twig as a "Skewer." Several species of shrikes *(Laniidae* spp.) impale or skewer their prey on thorns or other sharp projections on trees or bushes (e.g., Lorenz, 1950). The thorn is not manipulated by the bird in any way, and the behavior is not true tool-using. A pair of fiscal shrikes *(Lanius collaris)* in Tanzania used the same skewer throughout a two-year period, frequently flying 40 yards or more past other trees to the favored twig (personal observations). This use of a thorn skewer appears to be species-specific (Lorenz, 1950).

b. Use of Spine or Twig as a Probe. The only bird known to use tools as a normal part of its daily feeding pattern is the Galapagos woodpecker finch *(Cactospiza pallida),* which probes insects from crannies in the bark by means of a cactus spine or twig held in the beak (Fig. 1). When an insect emerges, the bird drops the spine and seizes the prey with its beak (Lack, 1947; Bowman, 1961; Eibl-Eibesfeldt and Sielmann, 1962; Eibl-Eibesfeldt, 1963, 1965). The bird has been seen to reject twigs that were too short. In addition, one bird tried (unsuccessfully) to break off the

FIG. 1. Galapagos woodpecker finch using twig to probe insect from crevice in bark. (Drawing from photo by Eibl-Eibesfeldt.)

end of an excessively long probe, and another, having twice tried to insert the forked end of a twig into a cranny, broke it off at the fork and was thus able to use the implement successfully (Bowman, 1961). The bird may, therefore, be said to show the beginnings of object modification.

Eibl-Eibesfeldt (1963, 1965, 1967) obtained a young male Galapagos woodpecker finch which had been taken from the nest as a fledgling. He found that, although the bird showed manipulative interest in twigs and frequently poked about with them, when it was hungry and saw an insect in a hole it dropped the twig and tried to capture the prey with its beak, often without success. Eventually, however, during manipulative play, it learned the appropriate use of a twig as a probing tool.

c. Use of the Bark "Plug." During his courtship display, the male satin bowerbird *(Ptilonorhynchus violaceus)* uses a mixture of charcoal and saliva to "paint" the inside of his bower. While smearing the mixture onto the walls with the side of his beak he holds a small "plug" of bark fiber at the end of his beak and thus prevents the paint from dripping out. The behavior, which must be regarded as tool-using, is probably species-specific like the rest of the display (Bowman, 1961).

d. Use of Stones and Rocks. Birds make use of stones and rocks as "anvils" against which to break open food objects; and also as true tools, picking up stones to drop or throw on hard-shelled food.

i. Dropping or hitting food objects onto selected hard surfaces. This behavior is not, of course, true tool-using. There are a number of birds which drop food objects from the air to the ground below. Some gulls (Larinae) and crows *(Corvus)* often drop shellfish from a height, but usually it seems to be a matter of chance whether the prey falls on hard or soft ground (Hartley, 1964). However, the Pacific gull *L. pacificus* (Hartley, 1964), the lammergeier, *Gypaetus barbatus* (Huxley and Nicholson, 1963), the raven, *C. corax,* (Lorenz, personal communication), and the bald eagle, *Haliaeetus leucocephalus* (Bindner, personal communication) usually drop food objects onto rocky sites. The raven and the lammergeier drop bones in order to crack them and feed on the marrow. Favored dropping places of the lammergeier are littered with bone fragments (Huxley and Nicholson, 1963). Bald eagles have been observed flying into the air with tortoises or turtles, which they dropped from heights of several hundred feet over highways. The eagles then swooped down to feed on their prey. (Bindner, personal communication) On one occasion when the shell did not smash on impact, the eagle picked up the turtle in one foot and beat it time and again until, presumably, the shell was broken.

The song thrush *(Turdus philomelos)* and a bowerbird *(Scenepoeetes dentirostris)* have both evolved the habit of smashing snails against rock "anvils" (e.g., Hartley, 1964). In addition, two other species make use of anvils in order to crack egg shells. The Egyptian vulture *(Neophron percnopterus)* normally breaks any egg which it can pick up in its beak by throwing it at the ground or another egg. One individual, however, was tested with plaster eggs. After trying unsuccessfully to break them in the usual way, it carried them to an anvil stone several yards away and threw them forcefully against that (van Lawick-Goodall and van Lawick, 1968). Two crested seriemas *(Cariama cristata)* tested in captivity (Kooij and van Zon, 1964), broke hens' eggs by dropping them onto the ground. When presented with a lime egg, both birds first dropped it as usual and then carried it to an anvil and dropped it on that. When this was unsuccessful a definite throwing movement, similar to that shown by the Egyptian vulture, appeared for the first time.

ii. Dropping and throwing stones onto food objects, etc. The Australian black-breasted buzzard may drop rocks from the air onto the eggs of emus *(Dromaius novaehollandiae),* the Australian bustard *(Ardeotis australis),* or an Australian crane *(Grus* sp.) and then fly down to feed on the contents (Chisholm, 1954). A pair of African black kites *(Milvus migrans)* was also observed dropping stones. One bird picked up three stones in its talons, one after the other, flew some fifty yards with them, and dropped them at about 5-minute intervals: the second picked up and dropped two stones. All five stones were dropped from a height of about 60 feet. Unfortunately the observer, one of our assistants, did not immediately investigate: subsequently we found four of the five stones within an area of some five square yards at the bottom of a shallow grassy gully among some low bushes. We could find no indication as to why they had been dropped, but the behavior was probably connected with feeding, as the birds showed no aggressive diving or screaming during the performance.

In 1850, Africans in South-West Africa described how Egyptian vultures would fly above ostrich eggs, drop stones from their talons, and, having cracked the shells, feed on the contents of the eggs (Andersson, 1857). More recently we have observed Egyptian vultures throwing stones at ostrich eggs to crack the shells (van Lawick-Goodall and van Lawick, 1966, 1968). The vulture picks up a stone in its beak and throws it at an egg with a forceful downward movement of head and neck (Fig. 2). It continues throwing until the shell is cracked.

In view of the fact that the seriema breaks small eggs by throwing them down in the same way as the Egyptian vulture, some experiments were carried out at the Amsterdam zoo, on our behalf, by Dr. Dekker.

FIG. 2. Egyptian vulture about to throw stone at ostrich egg.

Two individuals were first presented with an ostrich egg: both birds, after pecking and clawing at the egg, picked up one stone each. These stones were dropped onto the ground quite far from the egg. During a second experiment some time later, one bird, after trying to pick up the egg (a rhea egg this time), picked up a stone and *threw* it at the ground. On the third testing neither bird picked up a stone, but on the fourth and last testing, when the birds were offered a goose egg, one seriema picked up a stone, carried it to a large "anvil" rock, and threw its stone at it. Further experiments are planned to try and ascertain whether these birds, in the wild, may in fact open rhea eggs in this manner.

iii. Use of stones as "weapons" and sticks for beating. Two captive bald eagles (one old bird of some 35 years and the other a 2- to 3-year-old taken as a fledgling from the nest) were observed using objects aggressively on a variety of occasions. Three times the old bird took a small rock in one foot and used this to smash crickets (twice) and one giant hairy scorpion. On a number of other occasions both birds, after first trying to use their feet but being prevented by the jesses with which they were secured, picked up stones in their beaks and threw them horizontally forward for distances of up to 24 inches at crickets (Fig. 3). Several times the crickets were killed by such throws. On no occasion was the victim eaten: once a cricket was picked up in the beak but immediately shaken violently to the ground as though it tasted unpleasant.

Both birds threw stones in the same manner at a tame Western gopher turtle—which was unharmed. Also each of the eagles once used a stick to "beat" this turtle. The stick was held in the beak, swung upward with a dorsolateral movement of head and neck, and brought down forcefully onto the objective. These movements were repeated until the turtle moved out of reach. In view of the fact that these eagles are known to feed on turtles in the wild (see Section d, i above), it is not clear exactly what these captive birds were trying to accomplish. Finally, the young eagle, when held "on the glove," frequently picked up the ring on its chain and threw it toward its owner. This continued until the bird hit him on the arm and thus obtained his attention, whereupon it immediately ceased the bombardment (Bindner, personal communication).

I have described these observations in detail because of the surprising number of different motor patterns adapted to the use of tools, and also because some of the incidents represent the only well-documented evidence of a bird apparently using an object as a weapon.

FIG. 3. Bald eagle throwing stone at turtle and cricket. (From original sketch by C. Bindner.)

e. Miscellaneous. Under this heading are two observations of tool-use in individual birds.

i. Use of bread as "bait." One green heron *(Butorides virescens)* repeatedly placed pieces of bread in the water where it was feeding and then caught and fed on the fish which came to nibble on the "bait." The heron did not eat the bread, and chased off other water birds which tried to do so (Lovell, 1957).

ii. Use of a "scoop." A captive cockatoo *(Kakatoeinae* sp.*),* when the water in its drinking container was low, regularly used a half walnut shell to scoop out the liquid. How the bird acquired this behavior was not known (Fyleman, 1936).

f. Discussion of Tool-Using Behavior in Birds. There has been little experimental work on tool-using behavior in birds, so that it is possible only to speculate on the ontogeny or evolution of a few of the patterns involved. The use of a skewer in the shrike, and probably of snail-smashing behavior in the song thrush, is species-specific. These behaviors almost certainly derive from the habit, shown by many birds, of hitting live prey against the substrate before eating it (Hartley, 1964).

The dropping of food objects from a height may have evolved from accidental release of the food while the birds tried to prize the shells open during flight—as has been suggested as an explanation of tortoise-dropping in the lammergeier (Hartley, 1964).

The Egyptian vulture throws stones at ostrich eggs using exactly the same movements as those seen when it throws a smaller egg at the ground, and it is from this pattern that the tool-using behavior undoubtedly derives. Moreover, when vultures were prevented from approaching an ostrich egg owing to fear of our car or the presence of higher-ranking birds, they invariably picked up stones and threw them repeatedly at the ground. This could perhaps be labeled as a displacement activity stimulated by the sight of the egg, as could also the stone dropping shown by the captive seriemas when they were presented with large eggs. If tool-using does *not* occur naturally in the seriema, it would be interesting to find out whether displacement stone dropping could lead to the use of a tool through trial-and-error learning.

Finally, while our present knowledge is too limited to suggest that object play may be important to the development of tool-use in some birds, it should nevertheless be mentioned that it is known to occur in several of the known bird tool-users. Eibl-Eibesfeldt (1967) describes how woodpecker finches frequently played with probes when satiated, poking insects out of holes only to push them back in to "hunt" again. Sometimes two birds stood one each side of a log and pushed a mealworm back and forth, one to the other, through a narrow tunnel. Eibl-Eibesfeldt does suggest, as we have seen, that it is during such manipulatory play that a young and inexperienced bird may learn to use twigs as tools in the feeding context. Lammergeiers (Huxley and Nicholson, 1963) have been observed to drop bones and then make lightening dives to seize them just before they hit the ground. Such activities might be repeated again and again before the birds finally smashed the food for eating. Finally it may be significant that for the bald eagle, the bird which shows the greatest variety of tool-using patterns to date, more varied object play has been recorded than is known to occur in other species. Thus it may play with turtles in the air in the same manner as the lammergeier with a bone. In addition, captive individuals frequently threw stones and other objects in the air and then pounced on them, and one was observed throwing a snake up with one foot, catching it as it fell to the ground, and throwing it up again. Moreover, in the wild, parent bald eagles take a variety of objects to their nests, possibly for the young to play with. Rocks and turtles are thrown up and pounced on in the nest in play. Other objects found in nests range from spoons and balls to women's bras and panties (Bindner, personal communication). We shall return to the importance of object manipulation in play when discussing the development of tool use in the solving of problems in primates.

III. TOOL-USING AND RELATED BEHAVIOR IN MAMMALS OTHER THAN PRIMATES

There are far fewer examples of tool-using behavior in this category of animal than among birds, and most of the examples to be given below refer to observations on individual mammals only.

1. Tool-Using and Related Behavior for Obtaining Food

a. *Static Rocks Used as Anvils.* A number of species of mongooses smash eggs by throwing or flinging them against hard surfaces. The dwarf mongoose, *Helogale undulata,* flings an egg between its hind leg at a rock or other hard surface: the marsh mongoose, *Atilax paludinosus* stands upright with an egg between its front paws and throws it at the ground (Dücker, 1957). The banded mongoose *Mungos mungo* may throw an egg (also glomerid millipedes) between its legs or down onto the ground (Dücker, 1957; Eisner and Davis, 1967). A tame white-tailed mongoose, *Ichneumia albicauda,* picked up an egg in its mouth, carried it to a rock, and dropped it. (J. Amet, personal communication).

b. *Stones Used in Breaking Open Food Objects.* The only non-primate mammal known to show true tool-using behavior as part of its everyday behavior is the sea otter, *Enhydra lutris.* This animal dives to the sea bed and brings to the surface a shellfish together with an anvil stone. Lying on its back in the water it bangs the food against the rock, using both front paws, until the shell is broken (Fisher, 1939; Murie, 1940; Kenyon, 1959; Hall and Schaller, 1964). The latter investigators, working in California, in the southern part of the otters' range, found during a 3-week study that approximately 50% of the observed food intake of these otters was obtained by the use of anvils. It appears, however, that farther north the otters feed on different species of shellfish which the adults are able to open with their teeth: there are no reports of adults in the wild using anvils in this area, although a female was seen carrying a stone which she did not use. Youngsters, however, have been seen using anvils throughout the range—it appears that, in the north, their teeth are not strong enough to open the shellfish.

One experiment on two young wild-caught banded mongooses suggests that these mammals, too, may occasionally use objects as tools in their natural state. As we have seen, these mongooses typically crack eggs by hurling them through their legs at a rock anvil. When presented with an ostrich egg both mongooses first tried to push it between their legs, which was obviously impossible since the egg was larger than they were (Fig. 4), and then, with almost no hesitation, both took up stones

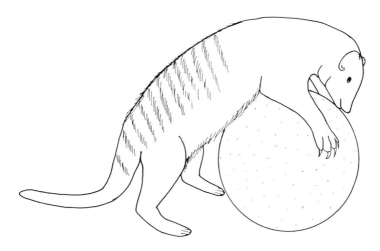

Fig. 4. Banded mongoose trying to hurl dummy ostrich egg between his legs. (Drawing from photo by P. McGinnis.)

in their front paws and hurled them back against the shell with good aim (Fig. 5). Whether or not they would have been successful is not certain, since the egg was a dummy (personal observation). Tests on wild mongooses of this species have not yet been successful because of their extreme wariness of strange objects.

2. Tools Used in Body Care

a. *Sticks Used for Scratching.* Three different nonprimate mammals have been observed to use sticks or straws for this purpose. Williams (1950) saw Burmese elephants pick up in their trunks long sticks with which to scratch their bodies. Huxley (cited in Thorpe, 1956) knew of a domestic goat which scratched itself with a straw held in its mouth, and Chapman (personal communication) has obtained evidence of a dock-tailed horse which frequently scratches itself with a stick (Fig. 6).

b. *Stone Used in Grooming.* Another tool-using individual is a Cocker Spaniel bitch which periodically uses a stone, marble, or other small

FIG. 5. Banded mongoose throwing stone between hind legs at plaster-filled ostrich egg.

hard object with which to "comb" the matted hair of her paws. She places the object against her upper incisors and draws the hair between the object and her tongue and/or lower incisors (Hart, personal communication; photos were sent as proof of this unusual behavior).

3. Objects Used as Missiles in an Aggressive Context

An Indian elephant in captivity, disturbed during the night by noises and lights in a neighboring cage, repeatedly hurled dung and straw over the dividing partition. This went on for several nights, the elephant gathering her "ammunition" into a pile in front of her before commencing the bombardment (Proske, 1957). To date the throwing of missiles has not been reported for wild elephants, although on two occasions African elephants *(Loxondonta africana)* tore off branches and waved them about in their trunks during mock charges or "intimidation displays" at the approach of a car (personal observation).

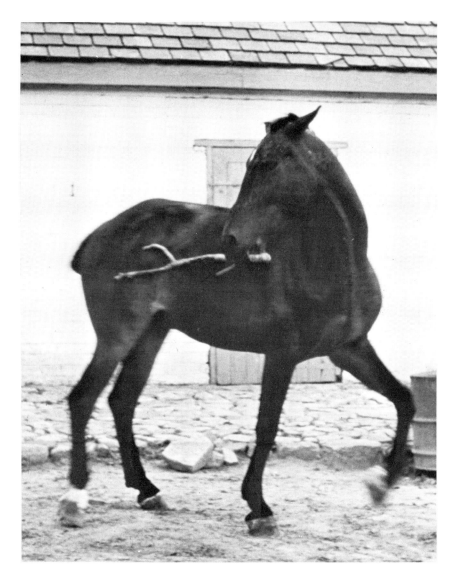

FIG. 6. Use of stick for scratching. (Photo by J. Chapman.)

4. Discussion of Tool-Use in Nonprimate Mammals

It appears, then, that so far as our present knowledge goes, only one nonprimate mammal, the sea otter, regularly uses objects as tools. So far as I am aware, no experimental work has been carried out on the development of the behavior. Hall and Schaller (1964) observed that young sea otters frequently manipulated objects, and sometimes pounded objects together during play. Since these manipulatory patterns were in no way stereotyped, Hall and Schaller suggest that they might provide the basis for learning the use of tools in the feeding situation. These investigators also speculate that young otters may learn the actual tool-using technique from their mothers during the long period of dependency (Hall and Schaller, 1964). However, it would seem that youngsters in the northern part of the range, where adults probably seldom (if ever) use anvils, would have little opportunity for direct observational learning.

The throwing of stones by the two banded mongooses may have derived from the mongoose habit of throwing eggs or other food objects against hard surfaces. It is important to determine whether this was a spontaneous response of two individuals, or whether it is, in fact, common practice among wild mongooses of this species.

IV. Tool-Use in Primates

Tool-using behavior in primates falls into two distinct categories; the use of objects as weapons in aggressive contexts; and in nonagonistic contexts for obtaining food, for investigation, and for body care. While the use of objects as missiles is fairly common in a number of monkeys as well as in apes, examples of other types of tool-use have been only rarely reported in monkeys with the exception of *Cebus* species.

It has been suggested (Hall, 1963; Jay, 1968) that investigations into tool-using by animals, particularly by primates, should be limited to wild populations. Captive primates, it is argued, are living under abnormal conditions and their basic repertory of behavior may thus be affected. Individuals may show patterns that are not normal to the species as a whole. Although this may be so, it can be argued also that the potential ability to manipulate objects as tools in a given species is a fact as important to our understanding of the evolution of tool-use in primates, including man, as is the knowledge of whether or not an animal uses tools in its wild state. The fact that the abnormal conditions of captivity may induce a new behavior suggests that such behavior could occur in the wild too, given the right environmental stimulus. And, quite apart from this, our knowledge of primate behavior in the natural state is still extremely limited: if a primate shows behavior in captivity which has not been observed in the wild, this by no means implies that it does *not* occur in the wild.

Therefore, I shall briefly discuss tool-using performances known to occur in captive primates (other than those actually *taught* by demonstrations) as well as those known to occur in wild groups: in fact, as will become apparent, tool-using is only infrequently recorded for captive individuals of species not known to use objects as tools in the wild.

A. OBJECTS WIELDED OR THROWN IN AGGRESSIVE CONTEXTS

It is possible to trace a gradual increase in the effectiveness of objects used by primates as "weapons" during aggressive interactions with predators, intruders, or conspecifics. At one end of the scale is the shower of twigs, fruits, and so forth that may be dropped by monkeys onto an intruder below: when a group of coati *(Nasua narica)* were treated to such a bombardment by a group of *Cebus* monkeys *(C. capucinus),* the "victims" merely waited under the tree to eat the ripe fruits included among the other "missiles" (Kaufmann, 1962). At the other end of the scale is the two-pound rock that may be hurled with deliberate aim by a chimpanzee: a research student at the Gombe Stream Research Centre, hit by such a missile, was limping for several hours and bruised for days.

Within the category of objects used during aggressive contexts, it is difficult to draw the line between true weapons, objects apparently purposefully used to intimidate predators or conspecifics, and objects used randomly during aggressive displays. Thus although a male chimpanzee who grabs a branch and waves and hurls it randomly during a charging display through his group cannot be said to be "using a weapon," his branch is, nevertheless, a more *effective* weapon than, for example, the handful of leaves that he may throw, with deliberate aim, toward a baboon. Nor is it always possible accurately to determine whether an object has been thrown or wielded randomly or purposefully. And so, particularly since the random use of objects may, as we shall see, lead directly to a more deliberate use of weapons, it is appropriate to discuss all aspects of object use that may occur in aggressive contexts rather than limit the field to weapon use as such.

1. Object-Use that Enhances Displays

A number of species of monkeys, gibbons, gorillas, and chimpanzees show characteristic displays, usually performed during intergroup encounters, but sometimes also as a response to intruders or during intragroup interactions. These displays are most frequently performed by adult males of the group. Vervets, *Cercopithecus aethiops* (Hall and Gartlan, 1965), patas monkeys, *Erythricebus patas* (Hall, 1965), colobus monkeys, *Colobus guereza* (Marler, 1968), and langurs, *Presbytis entellus* (Jay, 1968; Yoshiba, 1968) bounce and jump vigorously through the branches. During such displays quite large branches may be knocked off: in patas

(Hall, 1965) and langurs (Yoshiba, 1968) at any rate this often appears to be deliberate. Gibbons, *Hylobates lar* (Ellefson, 1968), swing through the trees and knock large dead branches down with their feet, also in a seemingly deliberate manner. The gorilla, *Gorilla gorilla beringei,* may perform a characteristic display that includes tearing off handfuls of vegetation and branches which may be hurled (usually randomly) as the animal charges (Emlen, 1962; Schaller, 1963). And the chimpanzee, *Pan trogolodytes* spp., during his branch waving or charging display may brandish and hurl large branches (Fig. 7) and throw or roll stones or large rocks (Reynolds and Reynolds, 1965; van Lawick-Goodall, 1968). In none of these displays are the branches or rocks typically directed toward other individuals; yet the crashing of branches through the leaves or the hurling of objects through the air undoubtedly serves to enhance the displays.

In the chimpanzee, such displays may sometimes function to maintain or enhance the social status of the individual performing it. One male chimpanzee at the Gombe National Park appeared to make deliberate use of abnormal objects to better his charging displays: this, in turn, probably led to his becoming the dominant male of the group. In 1964 he held a very low social status. In December that year he began to use empty 4-gallon parafin cans during his charging displays. Initially he used one can only, hitting it ahead of him with alternate hands or occasionally kicking it as he ran. After a while he was able to keep three cans on the move at once without noticeably diminishing his speed. The effect of such performances on his conspecifics was dramatic; the noise of the cans was tremendous and, as he approached, the other chimpanzees hurried out of his way, including those who held a much higher status. Often he repeated the display three or four times, running straight toward one or more of the other chimpanzees present. When he finally stopeed, the others usually approached and directed submissive gestures toward him. After 4 months we removed all cans, but by then he had acquired the number one position—which he still holds five years later. That his use of these cans was deliberate is suggested by the fact that, once the pattern was established, he would often walk calmly to the tent and select his cans. He dragged these quietly to a place from where he could, for instance, watch a group of conspecifics resting and then sat quite still for 5 minutes. Then he gradually started to rock, his hair slowly erected, he began a series of calls and then finally charged straight toward the peaceful group.

Chimpanzees and gorillas in captivity, deprived of natural display objects, may spit saliva or water, or throw feces at spectators, often after

FIG. 7. Mature male chimpanzee at the Gombe National Park brandishing a stick during a charging display.

rushing around the cage banging on the walls (Hewes, 1963; Riopelle, 1963; Wilson and Wilson, 1968; van Lawick-Goodall, personal observation).

2. *Objects Shaken or Dropped from Above*

This behavior occurs both in monkeys and apes.

a. In Monkeys. Some species of monkeys include in their threat reper-tory the vigorous shaking or hitting of branches (e.g., howler monkeys *Alouatta palliata*, Carpenter, 1934; red spider monkeys *Ateles geoffroyi*, Carpenter, 1935; rhesus monkeys *Macaca mulatta*, Hinde and Rowell, 1962; baboons *Papio* spp., Hall, 1962; DeVore and Hall, 1965; Hall and DeVore, 1965; van Lawick-Goodall, personal observation). This behavior may dislodge fruits and leaves and result in a shower of debris falling close to or onto the predator or other creature which has incited the monkey's aggression. In addition, in some of these species indi-viduals may purposefully break off branches, for example, and drop them with definite relation to the predator or intruder below. Howlers and red spider monkeys may react in this way to the presence of human observers (Carpenter, 1934, 1935) and, as we have seen, capuchins picked and dropped objects onto a group of coati (Kaufmann, 1962). No deliberate throwing movements were observed in any of these mon-keys although the spider monkey "may cause the object to fall away from the perpendicular by a sharp twist of its body or a swinging cir-cular movement of its powerful tail." Also these monkeys may pick branches and then retain them for a few moments until the intruder is more directly below (Carpenter, 1935).

Reliable reports of similar behavior in Old World monkeys made by trained observers are lacking, although Boulenger (1937) describes behavior that may be analogous in Patas monkeys *(Erythrocebus patas)* in West Africa, and Kortlandt and Kooij (1963) present undocumented instances of the dropping of objects by macaques, guenons, and colobus.

b. In the Apes. Gibbons (Carpenter, 1940), gorillas (Schaller, 1963), orangutans (Wallace, 1869; Schaller, 1961; Harrisson, 1962, 1963), and chimpanzees (personal observation) have all been observed to shake branches and break off and drop them in a similar way onto intruders below. The behavior seems to be particularly characteristic of the orang-utan, presumably because both chimpanzees and gorillas usually move away on the ground when disturbed, whereas orangutans frequently climb even higher in a tree and then drop or throw branches.

3. *Random and Aimed Throwing*

Throwing differs from the behavior described above in that a force-ful movement of the arm is involved: it may be apparently unaimed and random, or deliberately directed toward an objective.

a. In Monkeys. A captive *Cebus* monkey *(C. apella)* frequently threw objects at persons it "disliked." Initially it threw from the ground and was thus usually able to hit only people's legs. Eventually, however, it would climb quickly onto a chair or table with its missile, which it then threw at its victim's head (Romanes, 1882). Kortlandt and Kooij (1963) also cite reports of throwing in nine zoo *Cebus* monkeys.

Bolwig (1961) has recorded purposeful aimed throwing in a captive baboon, *Papio ursinus*. In addition Kortlandt and Kooij (1963) obtained reports on throwing in thirty-one baboons in zoos as compared with twelve for all other monkey species (excluding *Cebus*). As these authors themselves pointed out, the numbers of the different species in these zoos was not known and the figures, therefore, are of interest mainly because of the relatively large number of throwings recorded for baboons.

There is little evidence of true throwing in wild monkeys. Kortlandt and Kooij (1963) collected twelve reports of throwing in baboons, but no details of the motor patterns or behavioral contexts are given. It seems unlikely that the behavior is a common pattern among baboons in the wild: neither Hall nor DeVore (Hall, 1962; Hall and DeVore, 1965) observed throwing in the baboon groups they studied, nor has the behavior been seen in the baboons *(P. anubis)* at the Gombe National Park despite the fact that for five years they have been frequently aimed at and sometimes hit by stones thrown at them by chimpanzees (see below).

b. Throwing in Captive Apes. All three species of the great apes have been observed to throw objects in aggressive contexts in captivity (e.g., Köhler, 1925; Yerkes and Yerkes, 1929; Kortlandt and Kooij, 1963). Young chimpanzees, when first they arrive in captivity, usually show poor aim; the aim, however, improves with practice (Köhler, 1925; Kortlandt and Kooij, 1963). When young chimpanzees are taught to aim and throw experimentally, they may succeed in obtaining good scores (Morris, 1959; Kortlandt and Kooij, 1963).

c. Throwing in Wild Apes. As we have seen, gorillas and chimpanzees may throw branches, rocks, etc., at random during their chest-beating and charging displays. Orangutans *(Pongo pygmaeus)* also may hurl branches and so forth down toward intruders in a seemingly random manner (Schaller, 1961).

In addition to this random hurling, true aimed throwing has been observed in orangutans and chimpanzees in the wild. Orangutans may throw large fruits and branches in the direction of humans below (Schaller, 1961; Harrisson, 1962). Chimpanzee groups in the Congo and in

Guinea, when confronted with a stuffed leopard, were observed on a number of occasions to throw sticks toward the dummy: the groups living in thick rain forest showed poor aim whereas chimpanzees in the more open habitat in Guinea achieved a greater number of direct hits (Kortlandt, 1962, 1965, 1967; Kortlandt and Kooij, 1963).

In the Gombe Stream area, chimpanzees, on numerous occasions, aimed and threw sticks, stones, or other objects at baboons, humans, and occasionally conspecifics. Such behavior almost always occurred

FIGS. 8 and 9. Three and a half-year-old male chimpanzee throwing large stone at photographer.

when the context suggested that the throwing was aggressive. The chimpanzees threw either underhand or overarm (Figs. 8 and 9)—in either case the movements were similar to those made by a man performing similar actions. Kortlandt (1967) wrongly states that "the overhand throwing technique [does not] belong to the natural behavior inventory of the African apes." Throwing was more common in males than females (see also Kortlandt and Kooij, 1963), but one female at the Gombe became an expert thrower.

Although the data have not yet been fully analyzed, it is certain that, subsequent to the setting up of a feeding area for the chimpanzees (where they engage in frequent competition with baboons for bananas),

aimed throwing has increased both in frequency and efficiency.[2] Thus during the first year of the feeding station, five of sixteen male chimpanzees were observed to take aim and throw in aggressive contexts, and of the nine objects thrown only three were large enough to be potentially effective. By the end of the following year, aimed throwing had been recorded in three additional males and just over half of the thirty-two objects thrown that year were large rocks. The frequency of actual hits, however, showed no improvement (one the first year, four the second). The chimpanzees aim was good, but the missiles usually fell short of their objectives (Goodall, 1964; van Lawick-Goodall, 1968). Since then the frequency of throwing has continued to increase and is a fairly common aggressive pattern in some juveniles and one female. In addition a greater percentage of direct hits has been recorded.

4. Swaying Branches and "Whipping"

During an aggressive interaction, a chimpanzee may seize a growing branch or sapling and sway it very vigorously to and fro or up and down (van Lawick-Goodall, 1968). Frequently the "victim" is hit or whipped by the distal ends, sometimes very hard indeed. Kortlandt (1962, 1965, 1967; Kortlandt and Kooij, 1963) observed individuals in a number of different groups using saplings as whips in response to a stuffed leopard. At the Gombe Stream, the behavior was directed toward humans, baboons and, quite frequently, conspecifics, particularly during "dominance fights" between adult males (van Lawick-Goodall, 1968).

5. Branches Used for Brandishing, Hitting, and "Clubbing"

This type of behavior, although it has been observed in a monkey and occasionally in gorillas and orangutans, appears to be most common in captive and wild chimpanzees.

a. In a Monkey. There is only one well-documented observation of a monkey using a stick for hitting a victim. A *Cebus (Cebus fatuellus)*, after waving a stick around, used it to beat another monkey with which it was sharing a cage (Cooper and Harlow, 1961). Kortlandt and Kooij (1963) list a few records of "Agonistic (incipient) clubbing/stabbing" behavior in captive capuchins, and one for a baboon, but again they give no details of context, nor are the motor patterns described.

[2]A very similar increase in the frequency and efficiency of aimed throwing was reported by Köhler (1925) for his group of captive chimpanzees at Teneriffe.

b. In Captive Apes. Köhler (1925) briefly mentions one young female orangutan, who demonstrated "all shades of explosive actions with various implements up to complete armed attack on an enemy," and Harrisson (1963) describes how a young tame orangutan picked a twig, ran after a snake, and hit it. The other reports on the use of sticks as weapons in captive apes nearly all refer to chimpanzees. In a large group of chimpanzees in New Mexico several instances of one individual hitting another with a stick have been recorded. On one occasion the victim of a "gang fight" was beaten with a 2-foot long stick and received many cuts on his back (Wilson and Wilson, 1968). Kortlandt observed adult chimpanzees in an enclosure rush about brandishing large sticks when a live leopard was led to the top of the surrounding wall. Two adults rushed right up to the wall, dropping or flinging their sticks aside as they jumped up toward the leopard. On a later occasion, one of the adults, a mother, forcefully hit with a very large "club" the stuffed leopard which had been placed in the enclosure (Plate 8 in Kortlandt, 1967).

Kortlandt and Kooij (1963) also cite other examples of "clubbing" in captive chimpanzees. Again, however, the motor patterns are not described in detail, and it appears that these authors do not always differentiate between investigatory tapping or poking and aggressive hitting or stabbing. Thus they cite one instance of a chimpanzee which "alternated his striking at a pangolin by a few stabbing movements." This observation, in fact, concerned a chimpanzee that was merely investigating—tapping with his stick and occasionally poking at the strange animal (H. van Lawick, personal communication).

Köhler (1925) comments that the chimpanzees of his group at Teneriffe nearly always used sticks to investigate small creatures such as lizards or mice. But when the creature made a rapid movement in the direction of the chimpanzee concerned, the stick "became a weapon" and the victim was hit hard. If the creature did not manage to escape, it was killed anyway—not "deliberately" but "in the sheer excitement of the pursuit and capture." Kortlandt and Kooij (1963) also describe chimpanzees in captivity beating small creatures to death with sticks.

c. In Wild Apes. Gorillas may brandish branches during the chest beating display (Emlen, 1962; Schaller, 1963). One young male gorilla, holding a long stick in his hand, charged at a trapper. Before reaching his objective, however, he stopped, dropped the stick, and retreated (Kortlandt and Kooij, 1963).

Wild chimpanzees, as we have seen, frequently drag or wave branches during their charging displays. These may be picked up from the ground or torn from a tree (van Lawick-Goodall, 1968). In addition, while charging directly toward or past predators or other creatures which have aroused their hostility, chimpanzees have been observed to seize and brandish large sticks and branches. Thus many of the wild chimpanzees which encountered Kortlandt's stuffed leopard responded by waving and brandishing branches as they charged toward or around the model (e.g., Kortlandt, 1967).

At the Gombe National Park the chimpanzees sometimes brandished sticks in this manner during aggressive interactions with baboons (Fig. 10) and conspecifics. In addition these chimpanzees occasionally made purposeful hitting movements toward animals or other objects—such as their own reflections in a mirror (Fig. 11) and, once, some sort of insect (Fig. 12). The chimpanzee grasps the stick firmly in one hand and brings it down toward the objective. Sometimes these "clubbing" movements were very forceful, but quite often the weapon was released

Fig. 10.·Three and a half-year-old male brandishing a stick as he approaches baboons at the feeding area.

Fig. 11. Male chimpanzee just about to hit toward his mirror image.

before the moment of contact. Nevertheless, the behavior differed markedly from the random brandishing of branches described above. Once an old female received a very hard blow across her back, and I saw one baboon hit very hard with a dead palm frond. When chimpanzees used sticks as true weapons, they nearly always adopted a bipedal position (see also Kortlandt and Kooij, 1963; Wilson and Wilson, 1968).

Fig. 12. Infant "clubbing" an insect. The type of grip, at the moment of impact, is typical for these chimpanzees.

6. Use of Potential Weapons during Social Play

Just as aggressive patterns such as biting and hitting are incorporated by primates into their social play, so, in chimpanzees at any rate, we find examples of the playful use of "weapons." Köhler (1925) describes how the chimpanzees of his captive group frequently approached humans or conspecifics while brandishing sticks, and sometimes two would wave sticks at each other. Occasionally these chimpanzees hit each other quite hard during play, without causing an aggressive response from the playmate. It is significant that, if a quarrel did arise between two chimpanzees playing with sticks, then the "weapons" were flung aside and the animals attacked each other with hands and teeth. Köhler also observed the throwing of objects during play; and the use of pointed sticks for jabbing unsuspecting humans, dogs, or domestic hens also appeared to be a type of "game."

A similar use of potential weapons was also observed at the Gombe Stream. Occasionally large sticks were waved or dragged about during social play; the wild chimpanzees were not, however, observed to hit each other during play with sticks. Playful throwing was observed twice, both times during play bouts between juveniles: dead sticks were broken from a tree and thrown, overarm, with deliberate aim. Once an infant picked up a hard clod of earth to which was attached a few long dry grasses. Holding the ends of the grasses in one hand he swung the clod beating his playmate with it time and again.

7. Development and Evolution of Weapon-Use in Primates

As we have seen, many species of monkeys and all the apes may hit at or shake branches during aggressive displays directed toward enemies. As Hall (1963) has pointed out, it seems logical to suppose that if a branch shaken down by a monkey actually hits the intruder below, this will be more rewarding to the monkey concerned than if this did not happen. In the same way, a large branch which hits the enemy will probably produce a more rewarding response than would a small twig. It seems probable, therefore, that in species known for their learning ability, repeated experiences rewarded in this way might cause the slight modification of the threat-gesture repertory which is necessary in order for them to break off objects and drop them purposefully.

Tree-dwelling monkeys and the more arboreal apes have little incentive to develop true throwing behavior—it is often more practical to drop something through an observed gap in a tangle of branches than to try to throw. However, the fact that capuchin monkeys as well as orangutans are able to throw in captivity shows that arboreal species may indeed develop true throwing behavior.

That throwing behavior has been recorded in baboons is not surprising. As Hall (1963) pointed out, not only do baboons hold and shake branches in threat directed at an enemy, but they also manipulate stones as they turn them over during feeding and, when suddenly startled by a small noxious insect, these primates may make a swift, underarm hitting away movement. Thus the baboon has available the motor patterns necessary to direct objects as well as gestures toward predators or intruders.

The apes are anatomically better adapted to make throwing movements than are the monkeys: the ape shoulder girdle, like that of man, enables him to throw with power (Washburn and Jay, 1967), and he is also anatomically adapted for standing upright, a good posture for forceful throwing. In addition the chimpanzee, when threatening a conspecific, baboon, or human, may make arm movements very similar, or exactly similar, to those seen during aimed throwing (Fig. 13). On some occasions at the Gombe Stream, a chimpanzee let fly (almost by accident it seemed) some object that it happened to be holding—such as a banana—during an aggressive encounter with a baboon. Because the threat gesture was directed toward the baboon, the object also traveled in that direction.

FIG. 13. Juvenile chimpanzee directing threat gesture toward baboon. Compare Figs. 8 and 9. (Drawing from photos by H. van Lawick.)

Some chimpanzees at the Gombe Stream threw far more frequently than others: there are a few adult males who have never been observed to throw in aggressive encounters. Possibly, therefore, an important factor in the development of aimed throwing lies in the experience of the individual: the success of an "accidental" throw might well reinforce the behavior so that it is repeated in a similar aggressive context. Observational learning, which will be discussed in a subsequent section, might also play a role in the development of aggressive throwing.

In addition, throwing, as we have seen, occurs during the lone play of infants in the wild. This is not directed, but it provides an opportunity for the youngster to become familiar with stones and probably also aids in the development of the motor patterns involved. Throwing during social play in young chimpanzees, although it was rarely observed at the Gombe Stream, might also play a role in the development of aggressive aimed throwing.

A similar hypothesis may be put forward regarding the development of the use of sticks as weapons in chimpanzees. It is but a short step from the violent shaking of a branch to the deliberate swaying of a branch which causes the ends to touch or whip the object which has aroused the individual's hostility. Similarly, the random waving of branches during charging displays frequently causes other chimpanzees or baboons to rush out of the way — sometimes they may actually be hit. The screaming, running away, or cringing responses induced by such random branch-waving might well be sufficient "reward" to reinforce the branch-waving pattern. Moreover, some movements shown by a chimpanzee during normal threat or attack are very similar to the hitting movements when a stick is wielded.

Kortlandt's investigations with captive and wild apes have demonstrated convincingly the readiness with which chimpanzees will throw objects and brandish sticks in aggressive contexts. This has led Kortlandt to put forward a hypothesis which postulates that these behaviors must have had their origins in savanna-dwelling populations of chimpanzees and gorillas. When early man invented the spear, so Kortlandt argues, the apes, with only their primitive stick clubs and ability to throw, were driven back into the forests. In the more open forests, the chimpanzees retained their throwing and clubbing abilities to a considerable degree of proficiency: in the thick rain forests such activities were impractical and survived only as vestigial patterns. (e.g., Kortlandt, 1967).

An alternative theory to Kortlandt's "dehumanization" theory is that the anatomical structure of the chimpanzee, its inherent threat-gesture repertory, its innate tendency to manipulate objects (Schiller, 1949), and its marked ability to learn from past experience (e.g., Köhler, 1925; Yerkes, 1943) enables throwing and hitting to develop independently in different individuals.

That chimpanzees in more open habitats show a greater proficiency in throwing is not surprising, for they have a better chance to "exercise" and develop the behaviors. It is true that even in the thickest rain forest a chimpanzee could throw branches or brandish them as he displayed along an animal trail, but he certainly has less opportunity to acquire weapon use. It is equally plausible to regard weapon use as a latent behavior in forest populations of chimpanzees as to regard it as a vestigial pattern. Many people cannot play tennis, but everyone (if he is physically fit) has the capacity to learn how to.

Over and above these arguments is the fact that at least two forest-living primates, the capuchin and the orangutan, show true aimed throwing in captivity (Yerkes and Yerkes, 1929; Harrisson, 1963; Kortlandt and Kooij, 1963). Kortlandt argues that the zoo orangutan seldom throws as compared with the zoo chimpanzee (Kortlandt and Kooij, 1963). This, however, is possibly an artifact of zoo conditions, which are far more alien to the strictly arboreal orang than to the more terrestrial chimpanzee. Thus orangs usually become morose and lethargic very quickly: indeed, Harrisson (1963) writes that their development may be "arrested by lack of incentive to exercise." That such conditions are adverse to such behavior as spontaneous throwing is suggested by the fact that a young orangutan being reintroduced to the wild frequently threw, usually as a form of play but often with good aim (Harrisson, 1963).

Kortlandt (e.g., 1967) also postulates that chimpanzees use sticks as clubs during encounters with leopards in the wild. While it seems reasonable to suppose that chimpanzees might, in fact, use stones or sticks as missiles to *throw* at a predator, such as a leopard, it seems unlikely that, under normal circumstances, a chimpanzee would approach a healthy adult leopard close enough to hit it with a stick. Also, as Washburn (1963) has commented, "unless a stick is well selected and skillfully used an ape's teeth are far more effective." In Kortlandt's 1967 report he includes an excellent photograph of an adult female chimpanzee in a large enclosure hitting the stuffed leopard. She is using a very long stick which she is holding by its thinnest end. If the leopard were alive, this would scarcely be an effective weapon.

Kortlandt's hypothesis is, of course, based on the behavior he observed when he presented chimpanzees, in captivity with a chained, caged, or stuffed leopard; and in the wild, with the stuffed one. But it is not sensible to assume that an animal as "intelligent" as a chimpanzee is likely to be "duped" into believing that a stuffed leopard (even if it does wag its head) is normal, or that a chained or caged leopard is dangerous. Gnadeberg (1962) describes the behavior of a dog which, upon seeing that his stronger rival was tied up, calmly strolled into the other's territory, quite ignoring the vain struggles of the superior dog to pull free. "Evidently he recognized the lack of danger" comments the author. If a dog can take advantage of such a situation, then a chimpanzee most certainly can: one cannot, therefore, assume that the re-actions of chimpanzees to live wild leopards will be the same as those elicited by a stuffed model or a securely tied captive.

Unfortunately the only two encounters between chimpanzees and wild leopards which have been recorded did not involve mature male chimpanzees (Izawa and Itani, 1966; van Lawick-Goodall, 1968). In neither case did the chimpanzees involved use weapons of any type. It is necessary to await further evidence from the field before either rejecting or accepting Kortlandt's suggestion that "groups of savanna-dwelling chimpanzees do indeed fiercely attack living leopards in the wild" (Kortlandt, 1967).

B. TOOL-USING BY PRIMATES IN NONAGONISTIC CONTEXTS

The use of objects as tools in nonagonistic contexts has rarely been reported in free-living or captive primates with the exception of *Cebus* monkeys and the apes: only the chimpanzee is known to use tools fre-quently and in a wide variety of situations. In this section I shall discuss tool-using performances directed toward food-getting, investigation, and body care which occur in wild primates and in those living under normal captive conditions, together with a brief summary of perfor-mances which appear in experimental situations designed to test the ability of the subjects to use objects as tools in the solving of problems.

1. Tool-Using for Obtaining or Preparing Food (Other Than in Experimental Laboratory Situations

a. In Monkeys. Wild *Cebus* monkeys have been observed to peel the bark from twigs and then use them as probes to prize insects from under bark (Thorington, cited in Jay, 1968). In captivity *Cebus* monkeys (e.g., *C. apella, C. capuscinus*) have been observed on a number of occasions to use rocks or other hard objects to crack open nuts (Romanes, 1882; Vevers and Weiner, 1963; Tobias, 1965) and eggs (Kooij and Zon, 1964). Interestingly, these monkeys also open nuts by hitting them directly onto the ground, and an egg was opened in the same way. (Hill, 1960; Kooij and van Zon, 1964).

Rhesus macaques in Singapore have been observed using leaves to rub dirt from food—and from other objects such as an elastic band (Chiang, 1967). Possibly nonfood items are rubbed as a form of playful manipulative behavior.

The use of water by the Japanese macaque *(M. fuscata)* to wash sweet potatoes and wheat (Kawamura, 1959; Frisch, 1968) is not usually con-sidered tool-using, but it seems somewhat arbitrary to draw a line between the plucking of a leaf, which is invariably close to hand, and the deliberate immersion of a food object in water. Sometimes the monkeys travel a good many yards in order to wash their food: also, of course, the water cleans off the dirt far better than rubbing with leaves. A captive vervet monkey also developed the habit of washing food in her drinking bowl (Gartlan and Brain, 1968).[3]

In addition to the above, Kortlandt and Kooij (1963) mentioned a baboon squashing a scorpion with a stone and then eating it, and another using a stick to prod about in a termite nest. A *Colobus* monkey and a mangabey were also reported to use sticks while feeding on termites. It would not be surprising if these monkeys used tools in the wild: however, since no details of the behaviors, nor the conditions under which they were observed, are given, we should perhaps await further evidence before fully accepting these performances.

b. In the Great Apes. i. Use of rocks. The use of a stone as a hammer has been recorded only twice. Savage and Wyman (1843–1844) quoted an example of a chimpanzee using a rock to open a small hard-shelled fruit, and Beatty (1951) observed a chimpanzee in Liberia pound open the kernals of palm nuts in this way. One infant chimpanzee at the Gombe Stream pounded on the ground several times with a stone, but we were not able to find out what his objective had been—possibly the action was simply nondirected play.

ii. Use of sticks to reach food out of reach. While this is one of the most commonly discussed tool-using patterns of captive great apes, there are only two isolated observations of the behavior in the wild: one gorilla and one orangutan drew food toward them with sticks (Pitman, 1931; Kortlandt and Kooij, 1963).

iii. Use of sticks for digging. Köhler (1925) describes how the chimpanzees of his captive group ate roots by digging with their hands. The use of sticks for digging started independently of this behavior, as a form of play, but was quickly adapted to the digging for roots. The chimpanzees not only held the sticks in their hands: sometimes they placed the soles of their feet on the uppermost ends and pushed, and sometimes they gripped one end between their teeth and pushed with the head and neck. It is of interest that a young tame orangutan, when pushing a stick into an insect nest, also occasionally gripped the tool with his teeth (Harrisson, 1963).

[3]It may be of interest to note here that the so-called food "washing" behavior of the raccoon *(Procyon lotor)* does not appear to be motivated by a need to clean the food. The behavior may serve to create, in captivity, a natural situation and function "to allow the expression of a thwarted independent feeding mechanism" (i.e., the feeling for food objects in rivers) (Lyall-Watson, 1963).

iv. Use of sticks as levers. Captive chimpanzees and orangutans frequently use sticks, iron bars, or similar objects to try to force apart the bars or mesh of their cages (e.g., Boulenger, 1937; Harrison, 1962). One group spent long hours trying to lever up the lid of a water tank (Köhler, 1925). In a similar manner the chimpanzees at the Gombe Stream began to use sticks to try to prize open the lids of boxes containing bananas (Fig. 14). Usually the chimpanzee, after breaking off a suitable stick, stripped off the leaves (see also Köhler, 1925) and often

Fig. 14. Mature female trying to pry open lid of box that contains bananas. Her infant (on the box) and a juvenile watch intently.

bit splinters off one end so that they formed a chisel-shaped edge. The chimpanzees were very persistent in their attempts—presumably because we occasionally opened a box at which one was working and thus reinforced the behavior.

Another use of sticks as levers at the Gombe Stream was observed when two individuals made repeated attempts to force an opening into the large nest of a species of arboreal ant [*Crematogaster (Atopogyne) sp.*]. The chimpanzees tried to push their sticks between the nest and the branch to which it was attached. The walls of these nests are extremely hard and after some 5 minutes the chimpanzees abandoned their attempt.

v. Use of sticks, twigs, and grasses when feeding on honey and insects. Wild chimpanzees have been observed poking sticks into the nests of bees and licking off the honey (Merfield and Miller, 1956; Izawa and Itani, 1966). At the Gombe Stream, chimpanzees were observed using large sticks to enlarge the entrance of bees' nests: holding the sticks in both hands they pushed them backward and forward. They then reached for the honey with their hands.

At the Gombe Stream the chimpanzees also used sticks when feeding on two species of ant, the arboreal *Crematogaster,* as described above,[4] and *Anomma* sp. (safari ant). In both cases the sticks were pushed into the nests, left for a few seconds, and then withdrawn covered in ants which were then eaten either directly from the stick or after being gathered together as the chimpanzee swept the stick through his free hand. Occasionally, when a chimpanzee came across a line of *Anomma* traveling across its path, it held a grass stem or twig among the insects and picked off those that climbed onto the tool. This is exactly similar to ant-eating behavior observed by Köhler (1925) in all the individuals of his captive group.

The most frequently observed tool-using behavior in the Gombe Stream area is the use of stems and small twigs during termite *(Macrotermes bellicosus)* feeding. After opening up a passage in a termite mound, a chimpanzee picks a grass stem or small twig and pushes it carefully down the hole. After a slight pause he withdraws the tool and picks off with his lips and teeth the insects clinging to it. The tool is held between the thumb and the side of the index finger (Fig. 15) with a precision grip (Napier, 1960).

Some individuals use as a tool any material that is at hand; others carefully inspect various clumps of grass before selecting a tool. Often several tools are picked at a time for immediate and subsequent use. Tools are frequently prepared carefully: leaves may be stripped from twigs, strips pulled from blades of grass, thin strips of fiber detached from bark. When the end of a tool gets bent, the end may be bitten off

---

[4] I never found out whether the chimpanzees opened the nests themselves or took advantage of openings made by the ants during swarming.

FIG. 15. Old female "fishing" for termites.

(cf. Köhler, 1925 p. 94), the other end used, or a new tool selected. Sometimes when only a few termites, or none at all, are biting onto his tool, a chimpanzee may use one after another in quick succession, poking in two or three times with one tool and then discarding it for another — as though it is the tool that is at fault. No chimpanzee was observed to move out of sight of the nest at which it was working to collect a new tool, but often tools were selected for subsequent use when a nest might be as much as 100 yards away and out of sight (Fig. 16). One male traveled for half a mile with the same tool in his mouth, inspecting one termite nest after another, none of which was ready for working.

The use of tools for termiting was not observed in infants under two years, although from about 9 months of age youngsters sometimes watched their mothers intently (see also Fig. 14). Older infants, between one and two years old, often manipulated and prepared "tools" as a form of play activity during the termite season (three or four months a year). A one and a half-year-old once jabbed a short twig at the surface of a nest (there was no hole there) using the power grip (Napier, 1960) — rather as a small human infant holds a spoon or pencil (Gesell,

Fig. 16. Chimpanzee mother, after picking a grass tool for use at a termite nest approximately 60 yards away and out of sight, pulls her infant from a play session prior to leaving.

1940). Infants between two and three years of age used tools in the correct contexts, but the behavior was characterized by the use of inappropriate material (too short, too thick, etc.) and clumsy technique (the tool was often pulled from a hole with a jerking movement which would have dislodged any insect that had bitten on). Thus during twenty-two bouts observed in these infants (varying from a few seconds to 5 minutes in length), I saw only two termites caught. Three- to four-year-old infants still used tools inefficiently. The tools were sometimes longer, but they were often too flexible (see also Fig. 17, which shows a very flexible stem selected by an infant for pushing into an artificial "honey bowl"). In addition, during termiting, infants of this age often pushed only one or two inches of their tools into the hole—as compared with the five to eight inches often inserted by adults. Four-year-old infants showed a more adult technique, although their tools were usually shorter

Fig. 17. Three-year-old male infant using inappropriate tool (too long, too flexible) while sampling honey from an artificial underground "honey bowl" at the feeding area. The two adults who tried the honey selected long but very firm stems.

than those of adults; they often persisted for as long as 15 minutes as compared with the usual bouts of less than 5 minutes of the 3-year-olds.

vi. Use of stick to knock food object to the ground. This occurred once at the Gombe Stream when a mature male was afraid to take a banana held out to him by hand. After staring at the fruit he shook a clump of grasses in mild threat. He then shook them more violently and one of the stems touched the banana. He stopped shaking, let go of the grasses, plucked a thin plant from the ground, dropped it immediately and broke off a thicker stick. He then hit the banana to the ground, picked it up and ate it. When a second banana was held out he used the tool immediately. This observation is of interest since it was the only time when it was possible to observe what was probably an original solution to a completely new problem involving tool use in the wild.

   c. Use of Sticks, Twigs, and Grasses as "Olfactory Aids" during Feeding. During termite feeding behavior, a chimpanzee frequently pokes a stem into a hole which he has just opened and then, on pulling it out, sniffs intently at the end. Seemingly as a result of this behavior, he then either works at the hole or tries elsewhere. On four occasions young chimpanzees poked twigs into holes in rotten branches. After withdrawing their tools and sniffing the ends, three of them then broke open their branches. Twice this revealed grubs (probably wasp larvae) which were eaten. Once an adult wasp flew out. The fourth individual was only 3 years old: her "probe" was too thick and the end broke off in the hole. She tried to poke it out and then gave up. Another chimpanzee pushed a stick into a larger hole in a tree trunk and, after intently sniffing the end, dropped it and moved away.

   Three times, when I prevented a juvenile from reaching into my pocket to feel if there was a banana there, she poked long grasses in and then sniffed their ends. Each time there was, in fact, a fruit there and she followed me whimpering until I gave it to her.

   Other examples of sticks used to investigate occurred out of the feeding context and will be described under Section B,2 below.

   d. Use of Straws and Leaves for Drinking. In captivity chimpanzees have been observed dipping straws into water through cracks in a tank lid and licking them, and also drinking in this manner when water was fully accessible (Köhler, 1925). A chimpanzee at London Zoo dipped a straw into melted ice cream in a carton and licked it (personal observation). In the same zoo a gorilla selected an unbroken rye straw which he pushed through the bars until the crushed end reached a pool of his own urine outside. After soaking the absorbent end, the gorilla withdrew the stalk and put the liquid to his lips. This was repeated many times (R. Teleki, personal communication, 1968).

   The normal practice among chimpanzees at the Gombe Stream, when they were unable to reach water with their lips (e.g., when rainwater had collected in a natural water bowl in a tree trunk), was to use leaves as "sponges" to sop up the liquid. I saw individuals drinking in this way from natural hollows and also from bowls artifically scooped out at the feeding area. The chimpanzee usually strips leaves from a nearby twig and briefly crumples them by chewing, thus increasing their absorbancy. The leaf mass, held between the index and second finger, is then pushed into the bowl, withdrawn, and the water is sucked out.

A 2-year-old infant used leaves in the adult manner, but twice chose very tiny ones. A three and a half-year-old infant, however, after first dipping his hand into the water then poked in a piece of dry grass, using movements similar to those observed during termite feeding. Each time he put it to his mouth, he chewed the end: eventually, therefore, the stem was a tiny crumpled mass—a minute "sponge" (Fig. 18). Soon he abandoned this and poked in, again with termiting movements, a long narrow dead leaf. When he withdrew this, he immediately crumpled it in his mouth to make a sponge. On subsequent occasions he once used a similar leaf, but abandoned it after 2 minutes; once he used a leaf mass left by another individual; and once the back of his fingers again. Three other infants were observed using sponges in apparently playful contexts: the same may be true for the infant just described.

2. Sticks and Twigs Used to Investigate Unfamiliar Objects

In captivity chimpanzees readily use sticks to poke and prod at small creatures of which they are afraid or at fire (Köhler, 1925; Kortlandt and Kooij, 1963; Butler, 1965).

Among wild chimpanzees, I have not observed exactly this type of poking and prodding, but sticks are used to *smell* unfamiliar objects. When a dead python was placed at the feeding area one 8-year-old female, who had been staring at it for some time, first sniffed the end of a long palm frond on which the snake had lain, and then pushed it, hand over hand, until its tip touched the python's head (which was bloody). She then withdrew the implement and sniffed the end intently. (Pythons, although they occur in the area, are rarely seen.) On another occasion a juvenile, when he was prevented by his mother from touching his newborn sibling, repeatedly touched her gently with a stick and then sniffed the end (Fig. 19).

3. Objects Used in Body Care

Köhler (1925) comments on the fact that although the chimpanzees of his group often indulged in coprophagy, if they accidentally trod in feces they would limp away to find something with which to wipe themselves clean. Blood was dabbed with leaves or straw, often moistened with saliva, and one female constantly wiped her genital area with leaves during her menstrual periods.

FIG. 18. Three and a half-year-old male after crumpling a "grass stem tool" into a "sponge tool" while drinking at an artificial water bowl.

At the Gombe Stream the wild chimpanzees use leaves to wipe feces, mud, sticky fruit juice, urine, and blood from their bodies in precisely the same way (Fig. 20). I never observed menstrual blood being wiped away with leaves, but when a chimpanzee had diarrhea it often wiped

FIG. 19. Five-year-old "investigates" his newborn sibling.

itself clean with a very large handful of leaves. As Köhler also observed wounds were dabbed rather than wiped, and the leaves were then licked and reapplied to the wound. Several times chimpanzees pulled sprays of leaves toward them and rubbed vigorously after heavy rain. Once a juvenile wiped sticky banana from the head of her infant sibling with leaves. This was the only occasion when a chimpanzee was seen to wipe another individual; one infant, 9 months old, ran to his mother whimpering after falling into a heap of diarrhea, but she ignored the mess. No infant under 10 months of age was seen to wipe itself.

In addition to the above use of leaves, a female chimpanzee once used a twig apparently to pick her teeth, after first using her finger nail, and an infant briefly used a twig to pick his nose. No chimpanzee in the wild has, so far, been observed to use an object to scratch itself, as was observed in captive chimpanzees (Köhler, 1925).

4. Some Unusual Tool-Using Patterns That May Appear in Captive Primates

One common behavior in captive chimpanzees (Köhler, 1925; Yerkes, 1943) and orangutans (Harrisson, 1962, 1963) is the use of straw, cloth, or comparable materials as covering for the body at night, or during cold weather. One adult female chimpanzee, shut out in her open enclosure late in the evening, first started to make a nest but eventually collected a mass of leaves and straw and piled it all on her back (Köhler, 1925). Only one observation in the wild which may relate to this behavior con-

Fig. 20. Juvenile wipes blood from her clitoris with leaves after being bitten during a squabble.

cerns an adult male chimpanzee who picked a leafy branch and laid it over himself during rain (Izawa and Itani, 1966).

Another example of tool-use is the painting and drawing behavior described for captive capuchins, orangutans, gorillas, and chimpanzees (Morris, 1962; Rensch, 1965). Yerkes (1943) observed frequent use of feces by chimpanzees to plaster cage walls and other objects. Köhler (1925) gave chimpanzees lumps of white clay: after tasting the substance the chimpanzees wiped their lips on some object and showed interest in the white smears this made. Soon they began moistening clay with their mouths and "painting" objects with their whitened lips: eventually they used their hands to apply the clay rather like whitewash. A young orang-utan made marks with its finger in the wet sand (Harrison, 1962) and one 2-year-old wild chimpanzee at the Gombe Stream did the same on one occasion.

Finally one strange "tool-using" pattern has been recorded in three different primate species—the use of bread to feed animals of different species. A West African *Cercopithecus* sp. which was kept as a pet often sat above the household dog and held out pieces of bread for which the dog begged. Sometimes the monkey then dropped the food to the dog (personal observation). A tame capuchin, similarly a household pet, held out pieces of bread to lure ducks to come within its reach, which it then caught and killed (Boulenger, 1937). And a group of captive chimpanzees often held bread out to domestic hens and then, when the birds were within reach, stabbed them with sticks, apparently as a form of "amusement." One female actually fed the hens, holding the food in her hand as the hen pecked at it, or throwing bread outside the bars and "watching benignly" as the birds fed (Köhler, 1925).

It is not within the scope of this paper to discuss the many and varied ways in which objects are used as tools by home-raised apes (e.g., Kohts, 1935; Hayes, 1951).

5. Ability of Primates to Use Objects as Tools in Laboratory Experiments

Some monkeys, gibbons, and the great apes have been the subjects of laboratory experiments designed to reveal the extent to which they are capable of using tools to solve a variety of problems. Chimpanzees, in particular, have been extensively tested in this way.

a. Food out of Reach beyond Cage Bars. Capuchin monkeys *(Cebus* spp.) may use objects as varied as sticks, wire, rope, cardboard, and cloth to pull in food baits (Klüver, 1937). An individual chacma baboon *(P. ursinus)* used sticks to rake in food (Bolwig, 1961).

Yerkes tested the gibbon in this situation and found that it was able to pull in fruit with a rake provided the bait was between the cage and the rake at the start of the experiment. A gorilla was unsuccessful until the experimenter demonstrated the procedure. The orangutan was able to draw in food with a rake and also by throwing out and drawing in pieces of cloth (Yerkes and Yerkes, 1929). Chimpanzees are invariably able to solve this problem (e.g., Köhler, 1925; Yerkes and Yerkes, 1929; Birch, 1945; Schiller, 1949). Once a chimpanzee was reasonably ex-perienced in this test situation, Köhler (1925) observed that, although it occasionally picked up an unsuitable implement, it normally aban-doned this as it walked toward the bars and set off in search of more suitable material (e.g., a longer stick). One individual, however, held two short sticks in her hand in such a way that they "looked" like one long stick and repeatedly tried to use this inappropriate tool. Chimpanzees frequently showed object-modification or tool-making in this situation (see Section B,6 below).

  b. Food Objects Hung out of Reach. The chacma baboon mentioned above solved this problem by placing against the wall a stick, which she then used as a ladder (Bolwig, 1961). A *Cebus* monkey, *C. capucinus;* (Bierens de Haan, 1931), an orangutan (Yerkes and Yerkes, 1929), and chimpanzees (Köhler, 1925; Yerkes and Yerkes, 1929) successfully climbed sticks in this situation. The animal holds the stick upright under the food and then climbs up fast, seizing the reward as the stick falls. Cebus monkeys and chimpanzees are able to drag boxes underneath hanging food in order to reach it: one capuchin was able to stack three boxes in this way (Bierens de Haan, 1931), and chimpanzees may pile up one on top of the other up to five boxes (Köhler, 1925; Schiller, 1949). An orangutan and a gorilla, however, were able to use boxes in this way only after being taught (Yerkes and Yerkes, 1929). Chimpanzees (Köhler, 1925) and an orangutan (Yerkes and Yerkes, 1929) attempted to pull the experimenter under hanging fruit in order to climb him, and one chimpanzee tried to pull his conspecifics in the same way.

  c. Food Objects inside Boxes. Chimpanzees have successfully used long poles or rods to push food through a long, narrow box or tube open at both ends; this test involves the use of a tool to push the reward *away* from the animal in the first place (Yerkes, 1943; Khroustov, 1964). An orangutan solved this problem spontaneously, but a gorilla was unable to do so without being taught (Yerkes and Yerkes, 1929).

  Other problems involve the opening of closed boxes with keys, levers, screwdrivers, and so forth, but in most cases the use of the tool must first be taught and need not concern us here (e.g., Rensch and Döhl, 1967; Rensch and Dücker, 1966).

6. Object Modification or Toolmaking

  We may consider the seemingly simple act of breaking a branch from a tree as an initial step in the modification of an object for use as a tool. Laboratory chimpanzees unacquainted, or at least unfamiliar with, trees may not "perceive" an attached branch as a potential tool with which to reach food outside the bars. Schiller (1949) states that finally his chimpanzee subjects broke off branches, but that this was an expression of frustration which bore no relation to the problem in hand. Only when the branch was thus separated from the tree could it be perceived as a tool. Köhler (1925), however, whose group of chimpanzees was more familiar with trees, reported that some of his subjects broke off branches for tools with direct reference to the problem in hand, although most of them did not immediately come to this solution. In the instance of the wild chimpanzee male at the Gombe Stream who hit a banana from a human hand with a stick, the breaking off of the implement caused no greater difficulty, once he had found the solution to the problem, than would be the case with a man. It is of interest, too, that a wild-born baboon was also able to solve a problem that involved breaking a branch from a tree for use as a tool (Bolwig, 1961).

That it is, in fact, necessary for chimpanzees to become familiar with objects before they can use them as tools, let alone modify them in toolmaking, has been demonstrated by Birch (1945) and Schiller (1949). In addition Köhler (1925) and Schiller (1949) showed that certain complex manipulatory patterns involved in object modification (such as fitting two tubes together) could not be solved when the chimpanzee was "concentrating" on a food bait. Once the manipulation had been satisfactorily accomplished during free play, however, the pattern was normally available for use during a test situation. This fact should not surprise us: it is unlikely that an Australian aborigine who had lived all his life in the bush would immediately be able to use a pair of fire tongs to pick up a piece of coal which had fallen from the fire.

In the wild, of course, all the patterns used by the Gombe Stream chimpanzees in the modification of objects for tool-use — such as the stripping of leaves from twigs, the pulling of fibers from bark, or the crumpling of leaves for a "sponge" — are readily available. Such manipulations of twigs and leaves occur daily during feeding and nest-making (van Lawick-Goodall, 1968).

In captivity, extensive experiments have been conducted in order to find out to what extent the chimpanzee is capable of toolmaking. It has shown itself capable of breaking splinters or boards from boxes for use as "sticks"; uncoiling part of a length of wire for the same purpose; removing sand or rocks from boxes so that they may be pulled under a hanging bait; bending a few long thin straws in half to make a firmer "stick"; and fitting two and even three tubes together to form a long tool — provided, of course, it has become familiar with these objects prior to the test situation.

The most far-reaching experiments to date on the chimpanzee's capacity for toolmaking are those of Khroustov (1964) summarized by

Tobias (1965). A chimpanzee, having solved the simple problem of pushing a food bait through a hollow tube with a rod, was subsequently required to perform modifications of increasing complexity on a variety of materials in order to obtain a tool that could be inserted into the tube. The ape managed to break cross pieces from the ends of a stick, and to break suitable fragments first from rectangular and then circular boards of wood. After a series of experimental sessions, the above requirements were satisfactorily met. It was noted that the chimpanzee, when breaking off suitable splinters, followed the grain of the wood: when false grain was superimposed the ape initially tried to follow this, but subsequently broke off his tool along the true grain of the wood. As a final experiment the chimpanzee was presented with material which it could not break with its teeth or hands, together with a Chellean hand axe. At no time, however, did the chimpanzee attempt to use the implement provided, even though, for the first time in the series of experiments, he was repeatedly shown the correct use of the axe. Unless further experiments along these lines, perhaps with different materials or different individual chimpanzees, prove that a chimpanzee can, in fact, use a tool to make a tool, we must conclude that this stage of sophistication in toolmaking is beyond the mental capacity of apes.

7. Evolution and Development of Nonagonistic Tool-Use in Primates

It is of interest to consider the behavioral contexts from which this type of tool-using may be derived, and also the extent to which individual experience and learning play a part in the development of tool-using patterns.

a. Behavior from Which Tool-Using May Be Derived. The hands of living primates are well adapted to grasp and handle a variety of objects (Napier, 1960). Furthermore, monkeys and apes, both in the wild and in captivity, do manipulate things constantly, not only during feeding, grooming and so on, but also, particularly in the case of youngsters, during play and during exploratory behavior, when new or unusual features of the environment are carefully examined (e.g., Köhler, 1925; Yerkes and Yerkes, 1929; Menzel, 1964; Butler, 1965; van Lawick-Goodall, 1968). Schiller (1949) has suggested that many of the manipulative patterns observed in primates may be "innate" and that from such patterns adaptive behavior such as tool-using may be derived.

In captivity *Cebus* monkeys, as we have seen, may hammer open hard food objects against the ground or a rock. In addition, these monkeys frequently pick up stones and other objects and bang them against the ground, apparently as a form of play activity (Hill, 1960). Thus *Cebus* monkeys show a tendency to use objects as hammers from which purposeful tool-using may well be derived.

Young chimpanzees in captivity, if allowed to play with sticks, normally use these to poke or tap at a variety of objects before touching them with either hands or lips (Menzel, 1964; Butler, 1965). In addition chimpanzees, as well as other primates, may actually manipulate levers, buttons, and so on in test situations for no other reward than the performance of the activity (Schiller, 1949; Harlow, 1950). It is not difficult, then, to imagine that a young chimpanzee in the wild, as he played by a termite nest, might first of all scratch enquiringly at a spot of damp earth sealing a termite passage, and secondly poke a grass stem or twig into the hole which he thus revealed. Provided the youngster was familiar with termites (and all of the primates at the Gombe Stream feed on the winged forms of these insects) he would undoubtedly eat the insects which he found clinging to his "tool." Such a reward would undoubtedly induce him to push the twig once again into the hole.

Schiller (1949) found that young captive chimpanzees, once they had mastered a complex manipulatory pattern, frequently performed the actions, when not necessary, as a form of play. We have already seen that wild chimpanzee infants may prepare "tools" by stripping off leaves, shredding bark, and so forth, apparently as a form of play activity. In addition, a two and a half-year-old infant (i.e., soon after he had begun to use tools for termite-fishing in the correct context) was twice seen using grass tools out of context: once he pushed the stem twice through the hair of his own leg, each time touching the tool with his lips after withdrawal: once he pushed it carefully, three times, into another chimpanzee's groin. This "practising" of a tool-using pattern out of context is important, since it means that the behavior is readily available and may be adapted to new purposes. This, as we shall see below, may play an important role in the development of tool-using patterns in individual animals.

Thus it is suggested that the use of objects in connection with feeding may have evolved from a combination of certain investigatory and manipulatory behavior patterns which, in certain cases, became reinforced by food rewards.

b. Development of Tool-Using Behavior: Learning and Experience. Infant chimpanzees, both in the wild and in a variety of captive conditions, manipulate objects in a number of different ways during play and exploratory behavior, as do the other apes and many monkeys (e.g., Yerkes and Yerkes, 1929; Menzel, 1964, 1966; Butler, 1965). In many primates, including man, manipulative prowess gradually increases throughout childhood (e.g., Schiller, 1952; Mason *et al.*, 1959; Gesell, 1940) and Schiller (1952) found that the gradual increase in manipulatory skill of individual chimpanzees could be correlated with a gradual improvement in their ability to solve problems which entailed the use of objects as tools. This gradual improvement of manipulative ability is well demonstrated by wild chimpanzee infants in their handling of branches and twigs during feeding and nesting, as well as in tool-using situations (van Lawick-Goodall, 1968).

I have already suggested that a chimpanzee might develop a tool-using pattern, such as "fishing" for termites, from manipulatory patterns through trial-and-error learning. The repetition of such a pattern during infant play may, on occasion, lead to a new tool-using technique: for example, when the 3½-year-old infant used a grass stem, with the termite-fishing technique, to extract water from a hollow. As he sucked the moisture from the end, the grass became more and more crumpled until he had formed a "sponge." On another occasion a 4-year-old, when drinking with a "sponge," dropped it into the water bowl. She was unable to reach it with her fingers and, after a moment, picked a twig, stripped it of leaves and poked with it into the bowl. She touched the end to her lips, dropped it and then repeated the process with another twig. Her "purpose" in picking the twigs was unclear: nevertheless the availability of the pattern means that, in similar circumstances, it might be possible for a chimpanzee to use one tool not to *make* another, but at least to obtain a tool that is out of reach.

In addition to its ability to solve a problem through trial-and-error learning, the chimpanzee, and undoubtedly a number of other mammals as well, has evolved an ability that "supplements trial-and-error procedure by making possible forms of behavioural adaptation which strikingly resemble those which, in us, are known to depend upon perception of relation, ideation, insight or understanding" (Yerkes, 1943). The literature contains a number of descriptions of the sudden "purposeful" behavior of chimpanzees when they have, apparently through such "insight" or "ideation," perceived the solution to a test problem involving the use of objects as tools (e.g., Köhler, 1925, p. 23; Yerkes, 1943, pp. 135–136). A similar change in behavior, from the frustrated shaking of grasses to the sudden deliberate picking of a stick, was observed when the wild chimpanzee at the Gombe Stream solved the problem of how to obtain a banana from a human hand.

Finally we should enquire whether, over and above trial-and-error learning and occasional insight learning, tool-using patterns may be acquired, and thus transmitted from one generation to another, by means of observational learning. That some primates can, indeed, gain experience in the solution of novel problems through watching the actions of others has been shown experimentally (e.g., Darby and Rio-pelle, 1959; Warden *et al.*, 1940; Hayes and Hayes, 1952). It appears that the chimpanzee is actually able to acquire a *new motor pattern* by observation alone: the home-raised chimpanzee, Vicki, not only applied lipstick in the correct manner, but also pursed her lips whilst doing so, as she had seen Mrs. Hayes do on many occasions (Hayes and Hayes, 1952). This is one of the few observations which points to true imitation in mammals (Aronfreed, 1969).

In the wild, of course, infant chimpanzees have much opportunity for observing tool-use in adults. At the Gombe Stream, infant chimpanzees often watched adults intently as they used tools (e.g., Fig. 14) and sometimes, too, picked up and used tools that adults discarded. Twice a 3-year-old, after watching his mother wipe her bottom, picked leaves and did exactly the same himself. On neither occasion had he himself defecated, nor was there any sign of dirt on his bottom. In addition, as we have seen, infants too young to use tools for termiting, nevertheless began to prepare tools in a manner similar to adults, stripping off leaves and so forth, during the termiting season. Finally, one observation suggests that the adults may learn by observation: one female used a stick vigorously on a banana box the very first time she moved out into the feeding area. It seems most unlikely that she would have shown this response on a first encounter with boxes and bananas unless her behavior bore reference to the fact that she had watched, from the security of the surrounding trees, other chimpanzees behave in a similar way.

Bearing in mind the facts presented in the discussion above—the young chimpanzee's frequent investigations of his environment together with his innate tendency to manipulate and play with objects—it would seem reasonable to assume that chimpanzees probably show tool-using behavior of some sort or other throughout their range. This indeed is probably so, since examples of tool and/or weapon use have been gathered from Tanzania (in the eastern limit of the chimpanzee's range) and from Liberia (in the west), together with evidence also from other Central and West African countries (Savage and Wyman, 1843–1844; Merfield and Miller, 1956; Kortlandt, 1963; Izawa and Itani,

1966). If at least some tool-using patterns may be individually "dis-covered" and passed down through observational learning and imita-tion, then we should expect to find at least some different "cultural traditions" in chimpanzee groups that are geographically separated. Unfortunately our knowledge of tool-use for nonagonistic purposes, in areas other than the Gombe Stream, is not extensive: nevertheless the fact that in West Africa chimpanzees may use rocks as hammers, a behavior that has not yet been seen at the Gombe Stream, suggests that such, indeed, may be the case.

V. CONCLUDING REMARKS

There are tool-using invertebrates in the truest sense of the word, but there can be no question of such tool-using indicating "intelligent" adaptation. The larva of an ant-lion (Cicindelidae spp.) which flings grains of sand to knock struggling insects further into its funnel shaped pit is no more "gifted" than is the larva of a dragonfly (Anisoptera spp.) which has developed a hinged, elongated labium which it shoots out to grasp passing prey. One has evolved a behavioral mechanism which performs a similar function to the structural mechanism of the other.

It is when the motor patterns available to a given animal for the ma-nipulation of objects can be adapted to new situations that tool-using, in itself, becomes of special interest. The American bald eagle shows what must be considered, for a bird, a remarkable diversity of tool-using performances. Moreover, it is just possible that the throwing of stones and the wielding of sticks with the *beak* represented, initially, spontaneous adaptations by which the birds solved problems that they were unable to tackle in the normal way, i.e., with their feet. Such versa-tility has not been observed in other animals below the level of the pri-mates, although undoubtedly there are more tool-using animals and more tool-using patterns yet to be discovered.

Within the scope of existing knowledge it seems that, other than man, only the chimpanzee is able to adapt a variety of tool-using patterns to the solution of a rather wide variety of problems, both in captivity and in the wild. And this, it should be remarked, is not simply due to the anatomical structure of the chimpanzee: most of the higher primates have an opposable thumb and are capable of a type of precision grip (Napier, 1960), and all of them are certainly capable of grasping and poking around with a stick. In addition the ability to sit in an upright posture is widespread among the primates and this, as Tobias (1965) points out, is all that is needed for the hands to be freed for tool-use. It is undoubtedly a difference in the structure of the brain which dictates the frequency of tool-using performances in primates. Man, of course, has gone several stages further than the chimpanzee: he is able not only to use a tool to make a tool, but he makes tools to a set and regular pattern, he makes them for future use and for the use of others. Also, as Tobias (1965) and others have pointed out, man is the only animal dependent on tools for his survival.

A final point should be made. Once a species has evolved to the point where problems may be solved through individual experience including "insight," and when solutions may be transmitted to others by observational learning and immitation, it is necessary to emphasize the importance of *individual* performances. Just as there are exceptionally intelligent and exceptionally stupid humans, together with a vast majority that are of average capabilities, so it is with chimpanzees (and many other creatures too, for that matter). Given the possibility of techniques being learned by one animal from another, therefore, the presence of an exceptionally gifted individual in a free-ranging community may be of supreme importance in the development of tool-using cultures. This factor undoubtedly played a vital role in the appearance of tool-using and tool-making in early man, as it continues to do today.

Acknowledgments

The work on wild chimpanzee tool-using was carried out at the Gombe National Park (formerly Gombe Stream Chimpanzee Reserve) in Tanzania, East Africa; I am grateful to the Tanzania Government officials in Kigoma, the Tanzania Game Department personnel, and the Tanzania National Parks personnel, particularly the Director, John Owen, for their cooperation. The work was financed initially by the Wilkie Foundation and subsequently by the National Geographical Society and the Science Research Council, and I express my gratitude to all three organizations. In addition I also thank Dr. L. S. B. Leakey, who initiated this research, and the research assistants who have helped to carry on the work in the field.

The work on tool-using in Egyptian vultures was carried out in Ngorongoro Crater, and I express gratitude to the Conservator of the Ngorongoro Conservation Unit, Ole Saibul, and the Conservation Unit staff for their cooperation.

My gratitude is also due to Dr. Jacobi of the Amsterdam Zoo, who permitted us to carry out experiments on two crested seriemas, and, in particular, to Drs. Decker who made four experiments on these birds on our behalf. I also thank Mr. Douglas-Hamilton and Mrs. J. Amet, who helped us to test their tame mongooses.

My gratitude is due also to those who have written to me with information about tool-using in animals, in particular to Mr. J. Chapman, who not only wrote about a tool-using horse but tracked it down through two new owners and obtained, after much patience, photos and cine film of the behavior; and to Mr. Chuck Bindner, Jr., Specialist in Birds of Prey, Phoenix, Arizona, who not only informed me about tool-using behavior in bald eagles, but sent detailed notes for inclusion in this paper, including original drawings.

Finally my thanks are due to Baron Hugo van Lawick for his active participation in all aspects of the research related to tool-using behavior, and in the preparation of this review.

References

Andersson, C. J. 1857. "Lake Ngami or Explorations and Discovery during Four Years Wanderings in the Wilds of South Western Africa." Hurst & Becket, London.

Aronfreed, J. 1969. The problem of imitation. *In* "Advances in Child Development and Behavior" (L. P. Lipsitt and H. W. Reese, eds.), Vol. 4. Academic Press, New York.

Beatty, H. 1951. A note on the behavior of the chimpanzee. *J. Mammal.* **32**, 118.

Bierens de Haan, J. A. 1931. *Z. vergl. Physiol.* **13**, 639–695. [Cited in Hill (1960).]

Birch, H. G. 1945. The relation of previous experience to insightful problem-solving. *J. Comp. Psychol.* **38**, 367–383.

Bolwig, N. 1961. An intelligent tool-using baboon. *S. Afr. J. Sci.* **57**, 147–152.

Boulenger, E. G. 1937. "Apes and Monkeys." McBride, New York.

Bowman, R. I. 1961. Morphological differentiation and adaptation in the Galapagos finches. *Univ. Calif., Berkeley, Publ. Zoo..* **58**, 1–326.

Butler, R. A. 1965. Investigative behavior. *In* "Behavior of Nonhuman Primates" (A. M. Schrier, H. F. Harlow, and F. Stollnitz, eds.), Vol. 2. Academic Press, New York.

Carpenter, C. R. 1934. A field study of the behavior and social relations of howling monkeys. *Comp. Psychol. Monogr.* **10**(2).

Carpenter, C. R. 1935. Behavior of red spider monkeys in Panama. *J. Mammal.* **16**, 171–180.

Carpenter, C. R. 1940. A field study in Siam of the behavior and social relations of the gibbon. *Comp. Psychol. Monogr.* **16**, 38–206.

Chiang, M. 1967. Use of tools by wild macaque monkeys in Singapore. *Nature (London)* **214**, 1258–1259.

Chisholm, A. H. 1954. The use by birds of "tools" or "instruments." *Ibis* **96**, 380–383.

Cooper, L. R., and Harlow, H. F. 1961. Note on a Cebus monkey's use of a stick as a weapon. *Psychol. Rep.* **8**, 418.

Crook, J. H. 1960. Nest form and construction in certain West African weaver-birds. *Ibis* **102**, 1–25.

Darby, C. L., and Riopelle, A. J. 1959. Observational learning in the rhesus monkey. *J. Comp. Physiol. Psychol.* **52**, 94–98.

DeVore, I., and Hall, K. R. L. 1965. Baboon ecology. *In* "Primate Behavior" (I. DeVore, ed.). Holt, Rinehart & Winston, New York.

Dücker, G. 1957. Farb- und Helligkeitssehen und Instinkte bei Viverriden und Feliden. *Zool. Beitr. Berlin.* **3**, 25–99.

Eibl-Eibesfeldt, I. 1963. Werkzeuggebrauch beim Spechtfinken. *Natur Mus.* **93**(1).

Eibl-Eibesfeldt, I. 1965. *Cactospiza pallida* (Fringillidae). Werkzeuggebrauch beim Nahrungserwerb. *Encycl. Cinematogr.* **E 597/1964.**

Eibl-Eibesfeldt, I. 1967. Concepts of ethology and their significance in the study of human behavior. *In* "Early Behavior: Comparative and Developmental Approaches" (H. W. Stevenson, ed.). Wiley, New York.

Eibl-Eibesfeldt, I., and Sielmann, H. 1962. Beobachtungen am Spechtfinken *Cactospiza pallida* (Sclater und Salvin). *J. Ornithol.* **103**, 92–101.

Eisner, T., and Davis, J. A. 1967. Mongoose throwing and smashing millipedes. *Science* **155**, 577–579.

Ellefson, J. O. 1968. Territorial behavior in the common white-handed gibbon, *Hylobates lar*. (Linn). *In* "Primates: Studies in Adaptation and Variability" (P. Jay, ed.). Holt, Rinehart & Winston, New York.

Emlen, W. 1962. The display of the gorilla. *Proc. Amer. Phil. Soc.* **106**, 516–619.

Fisher, E. M. 1939. Habits of the Southern sea otter. *J. Mammal.* **20**, 21–36.

Frisch, J. 1968. Individual behavior and intertroop variability in Japanese macaques. *In* "Primates: Studies in Adaptation and Variability" (P. Jay, ed.). Holt, Rinehart & Winston, New York.

Fyleman, R. 1936. "Monkeys." Nelson, London.

Gartlan, J. S., and Brain, C. K. 1968. Ecology and social variability in *Cercopithecus aethiops* and *C. mitis*. *In* "Primates: Studies in Adaptation and Variability" (P. Jay, ed.). Holt, Rinehart & Winston, New York.

Gesell, A. 1940. "The First Five Years of Life: A Guide to the Study of the Pre-School Child." Methuen, London.

Gnadeberg, W. 1962. Erlebnisse mit Hunden. *Z. Tierpsychol.* **19**, 586–596.

Goodall, J. 1964. Tool using and aimed throwing in a community of free-living chimpanzees. *Nature (London)* **201**, 1264–1266.

Hall, K. R. L. 1962. Sexual, derived social, and agonistic behaviour patterns in the wild chacma baboon, *Papio ursinus*. *Proc. Zool. Soc. London* **139**, 284–327.

Hall, K. R. L. 1963. Tool-using performances as indicators of behavioural adaptability. *Curr. Anthropol.* **4**(5), 479–487.

Hall, K. R. L. 1965. Behaviour and ecology of the wild patas monkey, *Erythrocebus patas*, in Uganda. *J. Zool.* **148**, 15–87.

Hall, K. R. L., and DeVore, I. 1965. Baboon social behavior. *In* "Primate Behavior: Field Studies of Monkeys and Apes" (I. DeVore, ed.). Holt, Rinehart & Winston, New York.

Hall, K. R. L., and Gartland, J. S. 1965. Ecology and behaviour of the vervet monkey, *Cercopithecus aethiops*, Lolui Island, Lake Victoria. *Proc. Zool. Soc. London,* **145**, 37–56.

Hall, K. R. L., and Schaller, G. B. 1964. Tool-using behavior of the Californian sea otter. *J. Mammal.* **45**, 287–298.

Harlow, H. F. 1950. Learning and satiation of response in intrinsically motivated complex puzzle performance by monkeys. *J. Comp. Physiol. Psychol.* **43**, 289–294.

Harrisson, B. 1962. "Orang-utan." Collins, London.

Harrisson, B. 1963. Education to wild living of young orang-utans at Bako National Park, Sarawak. *Sarawak Mus. J.* **11**(21), 221–258.

Hartley, P. H. T. 1964. Article "Feeding habits." *In* "New Dictionary of Birds" (A. L. Thomson, ed.). Nelson, London and Edinburgh.

Hayes, C. 1951. "The Ape in our House. " Harper & Row, New York.

Hayes, K. J., and Hayes, C. 1952. Imitation in a home-raised chimpanzee. *J. Comp. Physiol. Psychol.* **45**, 450–459.

Hewes, G. W. 1963. Comments on Hall (1963).

Hill, W. C. O. 1960. "Primates," Vol. 4. Univ. of Edinburgh Press, Edinburgh.

Hinde, R. A., and Rowell, T. E. 1962. Communication by postures, gestures and facial expressions in the rhesus monkey *(Macaca mulatta)*. *Proc. Roy. Soc., Ser. B* **138**, 1–21.

Huxley, J., and Nicholson, E. M. 1963. Lammergeir, *Gypaetus barbatus,* breaking bones. *Ibis* **105**, 106–107.

Izawa, K., and Itani, J. 1966. Chimpanzees in Kasakati Basin, Tanzania. *Kyoto Univ. Afr. Stud.* **1**.

Jay, P. 1968. Primate field studies and human evolution. *In* "Primates: Studies in Adaptation and Variability" (P. Jay, ed.). Holt, Rinehart & Winston, New York.

Kaufmann, J. H. 1962. Ecology and social behavior of the coati, *Nasua narica*, on Barro Colorado Island, Panama. *Univ. Calif., Berkeley, Publ. Zool.* **60**, 95–222.

Kawamura, S. 1959. The process of subculture propagation among Japanese macaques. *J. Primatol. Primates* **2**(1), 43–60.

Kenyon, K. W. 1959. The sea otter. *Smithson. Inst. Annu. Rep.* **1958**, 339–407 [Cited in Hall and Schaller (1964).]

Khroustov, H. F. 1964. Formation and highest frontier of the implemental activity of anthropoids. *7th Int. Congr. Anthropol. Ethnol. Sci., Moscow.* [Cited in Tobias (1965).]

Klüver, H. 1933. "Behavior Mechanisms in Monkeys," Univ. of Chicago Press, Chicago, Illinois.

Klüver, H. 1937. Re-examination of implement-using behaviour in a Cebus monkey after an interval of three years. *Acta Psychol.* **2**, 347–397.

Köhler, W. 1925. "The Mentality of Apes." Harcourt, Brace, New York.

Kohts, N. 1935. Infant ape and human child. *Sci. Mem. Mus. Darwin, Moscow* **3**, 1–596.

Kooij, M., and van Zon, J. C. J. 1964. Gooiende Seriema's. *Artis* **6**, 197–201.

Kortlandt, A. 1962. Chimpanzees in the wild. *Sci. Amer.* **206**(5), 128–138.

Kortlandt, A. 1963. Bipedal armed fighting in chimpanzees. *Proc. 16th Int. Congr. Zool.* **3**, 64.

Kortlandt, A. 1965. How do chimpanzees use weapons when fighting leopards? *Year B. Amer. Phil. Soc.* pp. 327–332.

Kortlandt, A. 1967. Handgebrauch bei freilebenden Schimpansen *In* "Handegebrauch und verstandigung bei Affen und Fruhmenschen" (B. Rensch, ed.). Huber, Bern.

Kortlandt, A., and Kooij, M. 1963. Protohominid behaviour in primates (preliminary communication). *In* "The Primates" (J. Napier, ed.), Symp. Zool. Soc. London No. 10, pp. 61–88. Academic Press, London.

Lack, D. 1947. "Darwin's finches." Cambridge Univ. Press, London and New York.

Lancaster, J. B. 1968. On the evolution of tool-using behavior. *Amer. Anthropol.* **70**(1), 56–66.

Lorenz, K. Z. 1950. The comparative method of studying innate behaviour patterns. *Symp. Soc. Exp. Biol.* **4**, 221–268.

Lovell, H. B. 1957. Baiting of fish by a green heron. *Wilson Bull.* **70**, 280–281.

Lyall-Watson, M. 1963. A critical re-examination of food "washing" behaviour in the raccoon *(Procyon lotor* Linn). *Proc. Zool. Soc. London* **141**, 371–393.

Marler, P. 1968. Aggregation and dispersal: two functions in primate communication. *In* "Primates: Studies in Adaptation and Variability" (P. Jay, ed.). Holt, Rinehart & Winston, New York.

Marshall, A. J. 1960. Bower-birds. *Endeavour* **19**, 202–208.

Mason, W. A., Harlow, H. F., and Rueping, R. R. 1959. The development of manipulatory responses in the infant rhesus monkey. *J. Comp. Physiol. Psychol.* **52**, 555–558.

Menzel, E. W. 1964. Patterns of responsiveness in chimpanzees reared through infancy under conditions of environmental restriction. *Psychol. Forsch.* **27**, 337–365.

Menzel, E. W. 1966. Responsiveness to objects in free-ranging Japanese monkeys. *Behaviour* **26**, 130–150.

Merfield, F. G., and Miller, H. 1956. "Gorillas were my Neighbours." Longmans, London.

Morris, D. 1959. "The Great Apes." Granada Television Film.

Morris, D. 1962. "The Biology of Art." Knopf, New York.

Murie, O. J. 1940. Notes on the sea otter. *J. Mammal.* **21**, 119–131.

Napier, J. R. 1960. Studies of the hands of living primates. *Proc. Zool. Soc. London* **134**(4), 647–657.

Pitman, C. R. 1931. "A Game Warden among his Charges." Nisbet, London.

Proske, R. 1957. "My Turn Next: The Autobiography of an Animal Trainer." Museum Press, London.

Rensch, B. 1965. Uber ästhetische faktoren im erleben höherer Tiere. *Naturwiss. Med.* **2**(9), 43–57.

Rensch, B., and Döhl, J. 1967. Spontanes offnen vershiedener Kistenverschlüssedurch einen Schimpansen. *Z. Tierpsychol.* **24**, 467–489.

Rensch, B., and Dücker, G. 1966. Manipulierfähigkeit eines jungen Orang-utans und eines jungen Gorillas. Mit anmerkungen uber das Spielverhalten. *Z. Tierpsychol.* **23**, 874–892.

Reynolds, V., and Reynolds, F. 1965. Chimpanzees of the Budongo Forest. *In* "Primate Behavior" (I. DeVore, ed.). Holt, Rinehart & Winston, New York.

Riopelle, A. J. 1963. Comment on Hall (1963).

Romanes, A. 1882. Cited in Hill (1960).

Savage, T. S., and Wyman, J. 1843–1844. *Boston J. Natur. Hist.* **4**, 383.

Schaller, C. B. 1961. The orang-utan in Sarawak. *Zoologica (New York)* **46**, 73–82.

Schaller, C. B. 1963. "The Mountain Gorilla: Ecology and Behavior." Univ. of Chicago Press, Chicago, Illinois.

Schiller, P. H. 1949. Innate motor action as a basis of learning. *In* "Instinctive Behavior" (C. H. Schiller, ed.). Int. Univ. Press, New York.

Schiller, P. H. 1952. Innate consituents of complex responses in primates. *Psychol. Rev.* **59**, 177–191.

Thorpe, W. H. 1956. "Learning and Instinct in Animals." Methuen, London.

Tobias, P. V. 1965. *Australopithecus, Homo habilis,* tool-using and tool-making. *S. Afr. Archaeol. Bull.* **20**(80), Pt. IV, 167–192.

van Lawick-Goodall, J. 1968. The behaviour of free-living chimpanzees in the Gombe Stream Reserve. *Anim. Behav. Monogr.* **1**(3), 161–133.

van Lawick-Goodall, J., and van Lawick, H. 1966. Use of tools by the Egyptian vulture, *Neophron percnopterus. Nature (London)* **212**, 1468–1469.

van Lawick-Goodall, J., and van Lawick, H. 1968. Tool-using bird: the Egyptian vulture. *Nat. Geogr.* **133**(5), 631–641.

Vevers, G. M., and Weiner, J. S. 1963. Use of a tool by a captive Capuchin monkey *(Cebus fatuellus). Symp. Primate Biol., Zool. Soc., London.*

Wallace, A. 1869. "The Malay Archipelago." Macmillan, London.

Warden, C. J., Field, N. A., and Koch, A. M. 1940. Imitative behavior in Cebus and Rhesus monkeys. *J. Genet. Psychol.* **56**, 311–332.

Washburn, S. L. 1963. Comment on Hall (1963).

Washburn, S. L., and Jay, P. 1967. More on tool-use among primates. *Curr. Anthropol.* **8**, 253–254.

Williams, J. H. 1950. "Elephant Bill." Doubleday, New York.

Wilson, W. L., and Wilson, C. C. 1968. Aggressive interactions of captive chimpanzees living in a semi-free-ranging environment. Holloman Air Force Base Publ. ARL-TR-68-9.

Yerkes, R. M. 1943. "Chimpanzees, a Laboratory Colony." Yale Univ. Press, New Haven, Connecticut.

Yerkes, R. M., and Yerkes, A. W. 1929. "The Great Apes." Yale Univ. Press, New Haven, Connecticut.

Yoshiba, K. 1968. Local and intertroop variability in ecology and social behavior of Common Indian langurs. *In* "Primates: Studies in Adaptability and Variability" (P. Jay, ed.). Holt, Rinehart & Winston, New York.

Author Index

Numbers in italics refer to the pages on which the complete references are listed.

Subject Index

A

Adaptive radiation, 98–106
  parental care and, 103–105
    range of potential adaptations
    and, 98–102
Age,
  care of young and, 171–173
  voice recognition and, 54–56
Aggression,
  parent-offspring relationship and,
    effect on parents, 96
    effect on young, 84
  tool-using in,
    in birds, 201
    in nonprimate mammals, 206–207
    in primates, 209–225
Apes, *see* Primates

B

Behavior, *see also specific behaviors*
  homeostasis and, 9–15
    control theory and, 1–3
  ongoing, motivation and, 20–24
  parental, *see under* Mammalian con-
    specifics; Parent-offspring re-
    lationship
  social, in birds, *see* Voice recognition
Birds,
  tool-using in, 197–204
    in aggressive contexts, 201
    in obtaining and preparing food,
    197–201, 202
  voice recognition in, *see* Voice recog-
    nition
Body care,
  tool-using in, 205–206, 233–235

C

Care,
  of body, 205–206, 233–235
  of young, *see under* Mammalian con-
    specifics; Parent-offspring re-
    lationship
Carnivora,
  behavior toward young in,

    of females, 151–153
    of males, 132–135
  mother-infant recognition in, 167–168
  specificity of offspring in, 162–163
Cetacea,
  behavior toward young in,
    of females, 150–151
    of males, 131–132
  specificity of offspring in, 162
Chiroptera,
  behavior toward young in,
    of females, 140–142
    of males, 123
  mother-infant recognition in, 167
  specificity of offspring in, 160
Communication, parent-offspring relation-
    ship and,
    effect on parents, 96
    effect on young, 84–85
Conspecifics, *see* Mammalian conspecifics
Control theory, 3–4
  behavior and, 1–3
  block diagrams and, 8–9
  dynamic analogies to, 4–8

D

Dispersal, parent-offspring relationship
  and,
    effect on parents, 92–93
    effect on young, 79
Displays, tool-using in, 209–211

E

Energy, motivational, 16–20
Environment,
  animate,
    effect on parents, 93–98
    socialization of young and, 79–86
  physical,
    effect on parents, 90–93
    socialization of young and, 76–79
Experience,
  care of young and, 175–179
  tool-using and, 241–244

260